D0787707

FASTEN IT!

Other TAB Books by the Author

No. 1744
$23.95

FASTEN IT!

BY CHARLES R. SELF

TAB BOOKS Inc.
BLUE RIDGE SUMMIT, PA 17214

Dedicated with appreciation to Ruth and Sarah Honig, with a strong sidebar of thanks and affection to Ed and Marsha for having produced two such delightful children.

FIRST EDITION

FIRST PRINTING

Copyright © 1984 by TAB BOOKS Inc.
Printed in the United States of America

Library of Congress Cataloging in Publication Data

Self, Charles R.
Fasten it!

Includes index.
1. Joints (Engineering) 2. Fasteners. I. Title.
TA660.J64S45 1984   621.8′8   84-8887
ISBN 0-8306-0744-7
ISBN 0-8306-1744-2 (pbk.)

Cover photograph by the Ziegler Photography Studio of Waynesboro, PA.

Contents

Acknowledgments

With gratitude:

Any book on various modern processes has to be a cooperative effort by far more than the writer and the publisher. For that reason, I'd like to take time to express my gratitude to some of those who have helped gather material over the time this work has been in progress, and to apologize to those who—through oversight or other reasons—have been left out here.

The following are in no special order, and I'd very much like to emphasize that the interpretations and meanings read into their words and photographs are the author's own; in other words, if it's wrong, I goofed.

For long-term help, Bill Shanahan of The Stanley Works is invaluable, as is Vince Pax at Shopsmith. Don Cummings of Hyde Manufacturing and Robert Miller of Franklin Chemical Company have proved necessary to completion, and there are no real words to appreciate Jack Conover of Osborne Associates who handles informational details for The Cooper Group (companies such as Plumb, Weller, Xcelite, Crescent, Wiss, Lufkin, and Boker Tree).

Maze Nails has provided masses of information, as has Tom Hill at Hillwood Manufacturing. Lynda Nemeth is a late discovery at Black and Decker's Consumer Power Tools Division, but made up for my tardiness in location. Jim Leveque did an admirable job for

Black and Decker's Trade and Industrial Division.

Other companies providing assistance include Borden's—a lot!, Manco Tapes, Thermwell, Inc. (Frost King products, including duct and plastic tapes), 3M, Vaughan & Bushnell Mfg. (hammers, cold chisels and related items), Genova, Inc. (plastic plumbing fixtures, and Raingo gutter systems), through Rich Day, Homelite Division of Textron (and, yes, Virginia, you can use a chain saw to make rough wood joints), Sears, Roebuck & Co., DRI Industries, Vaco, Ridge Tool Company, Macco (part of Glidden Div. of SCM), and TECO Products and Testing Corporation, Chevy Chase, Maryland 20815.

All welding and brazing information is from Airco Welding Products of Murray Hill, NJ, and is much appreciated. Bostik provided hot glue, stapler, and drive-tool information that appears to be unavailable from other sources, with much detail on their Molly brand fasteners. Skil Corp. came through with some new power tool information.

I'd have to toss off the names of too many people to cover all those who have passed on tips and information over the years. These range from my father to the first foreman, who threatened the sanctity of my butt for miscutting a dado, and continue through today's friends who pass along the tips they know I need to make a living. To all those left out, my apologies, and maybe next time I'll remember.

Introduction

The field of fasteners is a wide one. Consider that polite society would be somewhat more rowdy without zippers or, for that matter, thread woven in a manner that stays fast. Our houses would collapse. Our vehicles would run worse than they already do on today's poor pretense for gasoline. Floorboards would constantly smack us in the face and dresser drawers would, instead of being hard to pull out, fall apart once out. Tents wouldn't stay up. Tool handles would have to be part and parcel of the tool (a goal some cheap claw hammers have reached, if it can be considered a goal). Books would be in loose leaf even sooner than provided for by modern production methods.

What is a fastener? I would almost say your guess is as good as mine. If some device, chemical, or design holds one or more items to one or more other items, it might be considered a fastener—and should be. We've got adhesives, nails, screws, bolts, welding, brazing, soldering, sewing, ropes, chains, mortars, cements, wood joints cut to fit and stay together, and clips and clamps and vises and . . . The list isn't endless but there is no lack of other fasteners.

The purpose of this book is to bring you all the common, and some not-so-common, types of fasteners of practical use around the home and farm. I will attempt to bring you the basics of wood joinery, both with and without glue in cut joints, and with nails, screws, and other fasteners.

Often there are three, four, or even more ways to fasten two objects together. Matching the fastener to the job required of the objects fastened becomes of great importance.

You must decide just what's to be done with the result. Will it need to be disassembled at some point? If that's so, then screws and nuts and bolts are probably the best consideration, although some forms of rivets are easily cut out and replaced. Must it resist no mechanical loading? Soldering may be the answer. Must it resist moderate to high stresses or shearing forces? Brazing could do the job, but if the stresses are great enough welding is required.

Clamping is normally a temporary expedient—temporary can sometimes be quite a long spread of months. Sheet-metal bends are better suited for lighter metals. The simple job of sticking a couple of metal plates together becomes more complex because today we have more fastening systems. Some glues and adhesives are superlatively strong. I'll probably fall into the habit of using the words glue and adhesive more or less interchangeably throughout this book (I'll try not to). It doesn't matter too much. Remember that glues are generally considered to be adhesives made of organic compounds. True adhesives can be all sorts of wild and wonderful chemical combinations that result in a bond stronger than the base material.

By the time you've finished this book, I hope you'll have a clear idea of just what you do need, and where you might like to go in continuing along the fascinating world of fastening things together.

Chapter 1

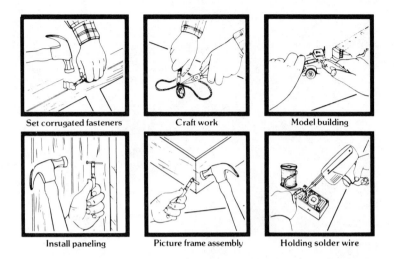

Set corrugated fasteners   Craft work   Model building

Install paneling   Picture frame assembly   Holding solder wire

Wood

Wood is the most common building material in our houses. It pays to know a bit more about it than most people do. Certain woods join easily with any glue, and withstand most extremes except heat (and some do pretty well there). Other woods cannot be successfully joined (Figs. 1-1 through 1-5) without special glues. Teak comes to mind because it is very oily.

Some woods nail easily; others are the devil's own when it comes to nail driving—seasoned locust, for a start—and are apt to split (white and red oak). All woods take screws well if pilot holes are drilled. Some woods are strong, some are weak, and most are available in a plethora of forms from solid to plywood, with stops at various veneers along the way.

New ways of using wood have developed over the years to increase utility, to cut price, and to add strength. Plywood provides great strength along the wood grain and gives you several plies with grain running in different directions. Plywood also allows the use of larger sections of solid wood than nature normally provides. Plywood veneer grains are bonded at right angles to adjacent layers, adding to strength.

Ways of fastening wood have changed and improved in many forms, though one could argue the beauty of old, handmade or machine made joints versus stainless steel screws and a glue line.

Fig. 1-1. Wood joints: dado (top); tongue & groove (center); rabbeted end lap (bottom). Courtesy of Shopsmith, Inc.

Fig. 1-2. Making finger joints in wood. Courtesy of Shopsmith, Inc.

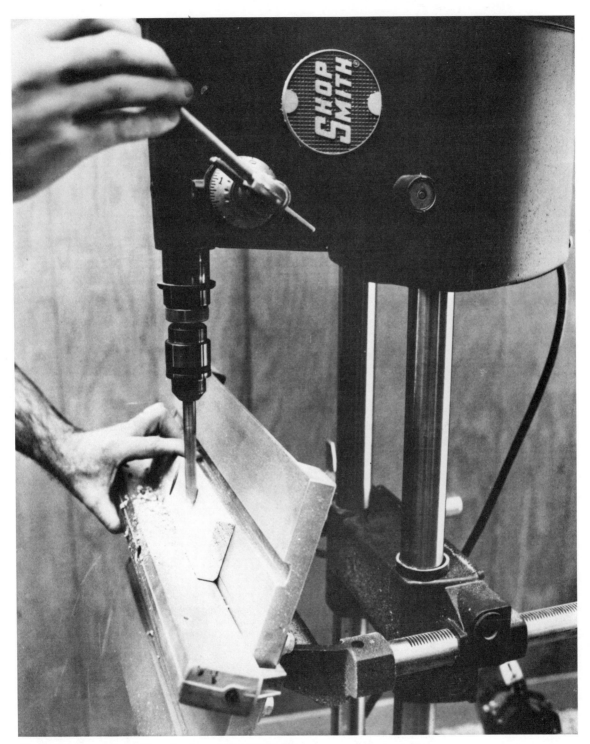

Fig. 1-3. Screw-pocket drilling. Courtesy of Shopsmith, Inc.

Still, structural adhesives used to put up panels in housing save time and prove as strong—sometimes stronger—than nailing.

The glues, or adhesives, used with exterior and interior plywoods are the primary difference in their durability and looks after time. Species across interior and exterior grades don't differ a

Fig. 1-4. A new way of cutting an old joint. Using a dado blade to make a dado. Courtesy of Shopsmith, Inc.

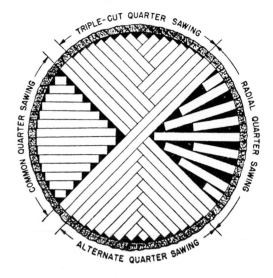

Fig. 1-5. Methods for quarter-sawing logs for less warping and cupping, and more attractive grain exposure.

great deal unless you're working with furniture like lumber-core plywoods or high-cost laminates. The species are pretty much dominated by the firs and pines.

Of course, the grade rating of the inner plies (C with most exterior and D with most interior) makes a strength difference in the overall panel that shows up structurally after the panels are fastened. If the glues won't withstand the climate, the strength of the plies has little meaning.

If only interior grade glues are used, the plywood must be subjected to extremes of humidity and dampness for extended time periods, and as near total dryness as possible is best. Interior plywood with intermediate glues (IMG) will be more resistant to molds, bacteria, and dampness, but it still cannot be allowed to stay damp or wet too long. Full exterior glue will provide the greatest protection and durability. Plywood surfaces will still check and wear from the weather if not protected with some form of finish or pressure treating.

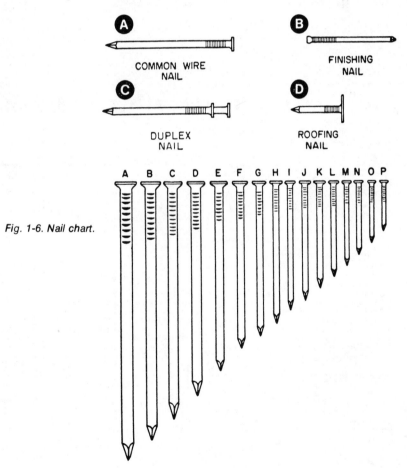

Fig. 1-6. Nail chart.

NAILS AND SCREWS – HOW TO USE THEM

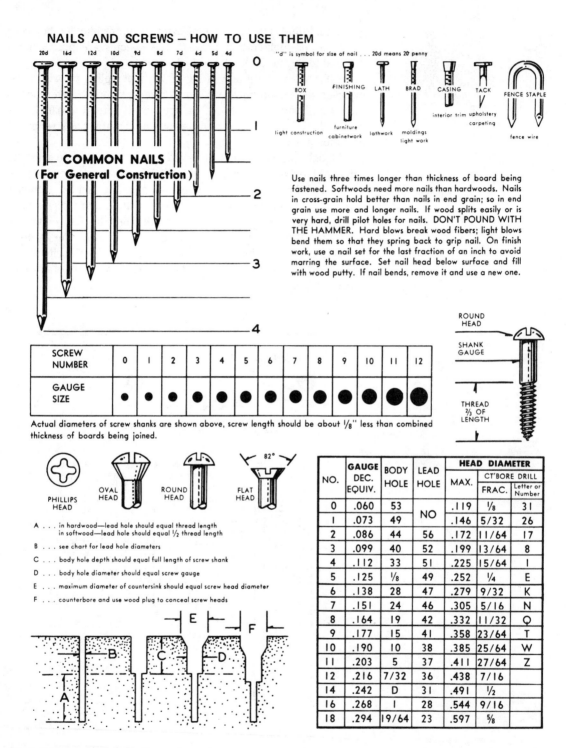

"d" is symbol for size of nail . . . 20d means 20 penny

COMMON NAILS (For General Construction)

20d 16d 12d 10d 9d 8d 7d 6d 5d 4d

BOX — light construction
FINISHING — furniture cabinetwork
LATH — lathwork
BRAD — moldings light work
CASING — interior trim
TACK — upholstery carpeting
FENCE STAPLE — fence wire

Use nails three times longer than thickness of board being fastened. Softwoods need more nails than hardwoods. Nails in cross-grain hold better than nails in end grain; so in end grain use more and longer nails. If wood splits easily or is very hard, drill pilot holes for nails. DON'T POUND WITH THE HAMMER. Hard blows break wood fibers; light blows bend them so that they spring back to grip nail. On finish work, use a nail set for the last fraction of an inch to avoid marring the surface. Set nail head below surface and fill with wood putty. If nail bends, remove it and use a new one.

ROUND HEAD
SHANK GAUGE
THREAD ⅔ OF LENGTH

SCREW NUMBER	0	1	2	3	4	5	6	7	8	9	10	11	12
GAUGE SIZE	●	●	●	●	●	●	●	●	●	●	●	●	●

Actual diameters of screw shanks are shown above, screw length should be about ⅛" less than combined thickness of boards being joined.

PHILLIPS HEAD
OVAL HEAD
ROUND HEAD
FLAT HEAD 82°

A . . . in hardwood—lead hole should equal thread length
        in softwood—lead hole should equal ½ thread length
B . . . see chart for lead hole diameters
C . . . body hole depth should equal full length of screw shank
D . . . body hole diameter should equal screw gauge
E . . . maximum diameter of countersink should equal screw head diameter
F . . . counterbore and use wood plug to conceal screw heads

NO.	GAUGE DEC. EQUIV.	BODY HOLE	LEAD HOLE	HEAD DIAMETER		
				MAX.	CT'BORE DRILL	
					FRAC.	Letter or Number
0	.060	53	NO	.119	⅛	31
1	.073	49		.146	5/32	26
2	.086	44	56	.172	11/64	17
3	.099	40	52	.199	13/64	8
4	.112	33	51	.225	15/64	1
5	.125	⅛	49	.252	¼	E
6	.138	28	47	.279	9/32	K
7	.151	24	46	.305	5/16	N
8	.164	19	42	.332	11/32	Q
9	.177	15	41	.358	23/64	T
10	.190	10	38	.385	25/64	W
11	.203	5	37	.411	27/64	Z
12	.216	7/32	36	.438	7/16	
14	.242	D	31	.491	½	
16	.268	1	28	.544	9/16	
18	.294	19/64	23	.597	⅝	

Fig. 1-7. Nail and screw chart.

Fig. 1-8. The Black & Decker Work-
mate serves as a vise and a work
center for small jobs and hobby
work. Courtesy of Black & Decker.

Hardwood plywoods are most often selected by face grades, although they may be bonded to either a lumber or hardboard core. Almost every wood with any sort of an attractive grain pattern is available either as plywood or as a laminate for you to apply.

Laminates are generally cut from 1/28th to 1/32nd of an inch thick, and are available with self-adhesive already applied, without

Fig. 1-9. Plywood end and edge
distance.

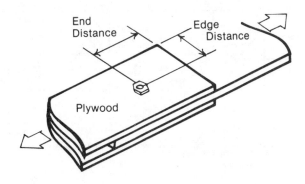

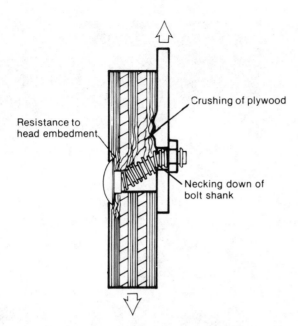

Fig. 1-10. Failure of a single-shear, plywood-and-metal connection.

Resistance to head embedment

Crushing of plywood

Necking down of bolt shank

adhesive, and in both stiff and flexible cuttings. Various sizes are available, with sheets commonly running up to about 24 inches by 96 inches. Beyond that you'll need to do your own matching at sides and ends. Small sheets and strips for edging are also available to provide a lower cost on small jobs and a finished look for all jobs. Edge finishing has long been a difficult task with many plywoods. Specially cut strips make it far easier than edge matching of laminates and contact gluing.

Now that I have described a little something about wood and its characteristics, I will move on to methods of fastening wood to wood and to other materials. There are three primary methods,

Fig. 1-11. Carriage bolts: plywood-and-metal connections.[a]

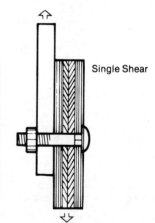

Single Shear

Bolt Diameter (in.)	Plywood Thickness (in.)	Plywood End Distance (in.)	Plywood Edge Distance (in.)	Ultimate Lateral Load (lb)[b]			
				Finger Tight		200-in.-lb Torque	
				Without Washer	With Washer	Without Washer	With Washer
3/8	1/2	1-1/2	3	1200	1475	1642	1829
	5/8	1-1/2	3	(1430)	(1710)	(1860)	(2230)
	3/4	1-1/2	3	1706	2520	2162	2698
1/2	1/2	2	3	1275	—	1556	—
	5/8	2	3	(1785)	—	(2320)	—
	3/4	2	3	1944	—	2711	—

(a) Plywood was C-D grade with exterior glue (all plies Group 1), face grain parallel to load. Side plate was 3/16"-thick steel. Bolts were No. 2 N.C. mild steel.

(b) Values in parentheses are estimates based on other tests.

Bolt Diameter (in.)	Plywood Thickness (in.)	Plywood End Distance (in.)	Plywood Edge Distance (in.)	Ultimate Lateral Load (lb)[b]			
				Single Shear		Double shear	
				Finger Tight	200 in.-lb Torque[c]	Finger Tight	200 in.-lb Torque[c]
1/4	1/2	1	3	1177	(1412)	967	(1160)
	5/8	1	3	1195	(1434)	1128	(1353)
	3/4	1	3	1504	(1805)	1377	(1652)
5/16	1/2	1-1/4	3	1490	(1788)	1427	(1712)
	5/8	1-1/4	3	1369	(1643)	1258	(1510)
	3/4	1-1/4	3	2040	(2448)	1772	(2126)
3/8	1/2	1-1/2	3	1729	2034	1514	1817
	5/8	1-1/2	3	1676	(2011)	1646	(1975)
	3/4	1-1/2	3	2120	2558	2328	2794
1/2	1/2	2	3	1870	2244	—	—
	5/8	2	3	(2090)	(2508)	—	—
	3/4	2	3	2236	2956	—	—

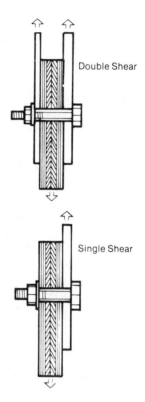

Double Shear

Single Shear

(a) Plywood was C-D grade with exterior glue (all plies Group 1), face grain parallel to load. Side plates were 3/16"-thick steel. Bolts were No. 2 N.C. mild steel, with washer.

(b) Values in parentheses are estimates based on other tests.

(c) Estimated values for 1/4" and 5/16" diameter bolts apply to torque level recommended by the bolt manufacturer.

Fig. 1-12. Machine bolts: Plywood-and-metal connections.[a]

Bolt Diameter (in.)	Plywood Thickness (in.)	Face Grain Direction	Plywood End Distance (in.)	Plywood Edge Distance (in.)	Ultimate Lateral Load (lb)
1/2	5/16	90°	4	2	2400
	1/2	0°	3	1	4740
		90°	3	1	4570
	5/8	0°	3	1-1/2	5400
		90°	3	1-1/2	4320
	3/4	0°	3	1-1/2	5280
		90°	3	3	5630
3/4	1/2	0°	4	2	7800
		90°	4	1-3/4	6900
	5/8	0°	4	1-1/2	8070
		90°	4	1-1/2	6900
	3/4	0°	6	1-1/2	7920
		90°	6	1-1/2	8650

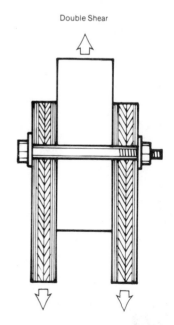

Double Shear

(a) Plywood was all-Group 1. Lumber was green stud-grade Douglas fir 2 x 4. Bolt grade was unknown.

Fig. 1-13. Bolts: plywood-to-lumber connections.[a]

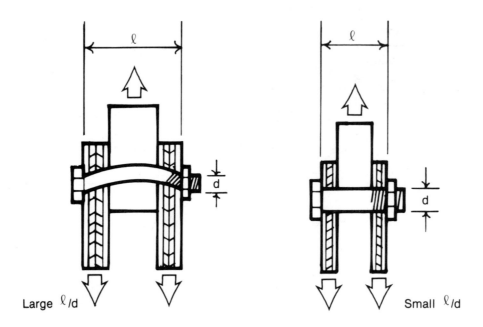

Fig. 1-14. Influence of bolt diameter on plywood-to-lumber connections.

Single Shear

Bolt Diameter (in.)	Plywood Thickness (in.)	Face Grain Direction	Plywood End Distance (in.)	Plywood Edge Distance (in.)	Ultimate Lateral Load (lb)	
					Without Washer	With Washer
1/2	1-1/8	90°	1-1/2	1-1/2	2003	2710
3/4	1-1/8	90°	1-1/2	1-1/2	—	3840

(a) Plywood was STURD-I-FLOOR 48 o.c. (2-4-1). Bolt grade was unknown.

Fig. 1-15. Bolts: plywood-to-plywood connections.[a]

Table 1-1. Common Woods. (Continues through page 14.)

Type	Sources	Uses	Characteristics
Ash	East of Rockies	Oars, boat thwarts, benches, gratings, hammer handles, cabinets, ball bats, wagon construction farm implements.	Strong, heavy, hard, tough, elastic, close straight grain, shrinks very little, takes excellent finish, lasts well.
Balsa	Ecuador	Rafts, food boxes, linings of refrigerators, life preservers, loud speakers, sound-proofing, air-conditioning devices, model airplane construction.	Lightest of all woods, very soft, strong for its weight, good heat insulating qualities, odorless.
Basswood	Eastern half of U.S. with exception of coastal regions.	Low-grade furniture, cheaply constructed buildings, interior finish, shelving, drawers, boxes, drainboards, woodenware, novelties, excelsior, general millwork.	Soft, very light, weak, brittle, not durable, shrinks considerably, inferior to poplar, but very uniform, works easily, takes screws and nails well and does not twist or warp.
Beech	East of Mississippi, Southeastern Canada.	Cabinetwork, imitation mahogany furniture, wood dowels, capping, boat trim, interior finish, tool handles, turnery, shoe lasts, carving, flooring.	Similar to birch but not so durable when exposed to weather, shrinks and checks considerably, close grain, light or dark red color.
Birch	East of Mississippi River and North of Gulf Coast States, Southeast Canada, Newfoundland.	Cabinetwork, imitation mahogany furniture, wood dowels, capping, boat trim, interior finish, tool handles, turnery, carving.	Hard, durable, fine grain even texture, heavy, stiff, strong, tough, takes high polish, works easily, forms excellent base for white enamel finish, but not durable when exposed. Heartwood is light to dark reddish brown in color.
Butternut	Southern Canada, Minnesota, Eastern U.S. as far south as Alabama and Florida.	Toys, altars, woodenware millwork, interior trim, furniture, boats, scientific instruments.	Very much like walnut in color but softer, not so soft as white pine and basswood, easy to work, coarse grained, fairly strong.
Cypress	Maryland to Texas, along Mississippi valley to Illinois.	Small boat planking, siding, shingles, sash, doors, tanks, silos, railway ties.	Many characteristics similar to white cedar. Water resistant qualities make it excellent for use as boat planking.

Type	Sources	Uses	Characteristics
Douglas Fir	Pacific Coast, British Columbia.	Deck planking on large ships, shores, strong-backs, plugs, filling pieces and bulkheads of small boats, building construction, dimension timber, plywood.	Excellent structural lumber, strong, easy to work, clear straight grained, soft, but brittle. Heartwood is durable in contact with ground, best structural timber of northwest.
Elm	States east of Colorado.	Agricultural implements, wheel-stock, boats, furniture, crossties, posts, poles.	Slippery, heavy, hard, tough, durable, difficult to split, not resistant to decay.
Hickory	Arkansas, Tennessee, Ohio, Kentucky.	Tools, handles, wagon stock, hoops, baskets, vehicles, wagon spokes.	Very heavy, hard, stronger and tougher than other native woods, but checks, shrinks, difficult to work, subject to decay and insect attack.
Lignum Vitae	Central America.	Block sheaves and pulleys, waterexposed shaft bearings of small boats and ships, tool handles, small turned articles, and mallet heads.	Dark greenish brown, unusually hard, close grained, very heavy, resinous, difficult to split and work, has soapy feeling.
Live Oak	Southern Atlantic and Gulf Coasts of U.S., Oregon, California.	Implements, wagons, ship building.	Very heavy, hard, tough, strong, durable, difficult to work, light brown or yellow sap wood nearly white.
Mahogany	Honduras, Mexico, Central America, Flordia, West Indies, Central Africa, other tropical sections.	Furniture, boats, decks, fixtures, interior trim in expensive homes, musical instruments.	Brown to red color, one of most useful of cabinet woods, hard, durable, does not split badly, open grained, takes beautiful finish when grain is filled but checks, swells, shrinks, warps slightly.
Maple	All states east of Colorado, Southern Canada.	Excellent furniture, high-grade floors, tool handles, ship construction cross-ties, counter tops, bowling pins.	Fine grained, grain often curly or "Bird's Eyes," heavy, tough, hard, strong, rather easy to work, but not durable. Heartwood is light brown, sap wood is nearly white.
Norway Pine	States bordering Great Lakes.	Dimension timber, masts, spars, piling, interior trim.	Light, fairly hard, strong, not durable in contact with ground.
Philippine Mahogany	Philippine Islands	Pleasure boats, medium-grade furniture, interior trim.	Not a true mahogany, shrinks, expands, splits, warps, but available in long, wide, clear boards.

Type	Sources	Uses	Characteristics
Poplar	Virginias, Tennessee, Kentucky, Mississippi Valley.	Low-grade furniture cheaply constructed buildings, interior finish, shelving, drawers, boxes.	Soft, cheap, obtainable in wide boards, warps, shrinks, rots easily, light, brittle, weak, but works easily and holds nails well, fine-textured.
Red Cedar	East of Colorado and north of Florida.	Mothproof chests, lining for linen closets, sills, and other uses similar to white cedar.	Very light, soft, weak, brittle, low shrinkage, great durability, fragrant scent, generally knotty, beautiful when finished in natural color, easily worked.
Red Oak	Virginias, Tennessee, Arkansas, Kentucky, Ohio, Missouri, Maryland.	Interior finish, furniture, cabinets, millwork, crossties when preserved.	Tends to warp, coarse grain, does not last well when exposed to weather, porous, easily impregnated with preservative, heavy, tough, strong.
Redwood	California.	General construction, tanks, paneling.	Inferior to yellow pine and fir in strength, shrinks and splits little, extremely soft, light, straight grained, very durable, exceptionally decay resistant.
Spruce	New York, New England, West Virginia, Central Canada, Great Lakes States, Idaho, Washington, Oregon.	Railway ties, resonance wood, poles, airplanes, oars, masts, spars, baskets.	Light, soft, low strength, fair durability, close grain, yellowish, sap wood indistinct.
Sugar Pine	California, Oregon.	Same as white pine.	Very light, soft, resembles white pine.
Teak	India, Burma, Siam, Java.	Deck planking, shaft logs for small boats.	Light brown color, strong, easily worked, durable, resistant to damage by moisture.
Walnut	Eastern half of U.S. except Southern Atlantic and Gulf Coasts, some in New Mexico, Arizona, California.	Expensive furniture, cabinets, interior woodwork, gun stocks, tool handles, airplane propellers, fine boats, musical instruments.	Fine cabinet wood, coarse grained but takes beautiful finish when pores closed with woodfiller, medium weight, hard, strong, easily worked, dark chocolate color, does not warp or check, brittle.

Type	Sources	Uses	Characteristics
White Cedar	Eastern Coast of U.S., and around Great Lakes.	Boat planking, railroad, ties, shingles, siding, posts, poles.	Soft, light weight, close grained, exceptionally durable when exposed to water, not strong enough for building construction, brittle, low shrinkage, fragment, generally knotty.
White Oak	Virginias, Tennessee, Arkansas, Kentucky, Ohio. Missouri, Maryland, Indiana.	Boat and ship stems, sternposts, knees, sheer strakes, fenders, capping, transoms, shaft logs, framing for buildings, strong furniture, tool handles, crossties, agricultural implements, fence posts.	Heavy, hard, strong, medium coarse grain, tough, dense most durable of hardwoods, elastic, rather easy to work, but shrinks and likely to check. Light brownish grey in color with reddish tinge, medullary rays are large and outstanding and present beautiful figures when quarter sawed, receives high polish.
White Pine	Minnesota, Wisconsin, Maine, Michigan, Idaho, Montana, Washington, Oregon, California.	Patterns, any interior job or exterior job that doesn't require maximum strength, window sash, interior trim, millwork, cabinets, cornices.	Easy to work, fine grain, free of knots, takes excellent finish, durable when exposed to water, expands when wet, shrinks when dry, soft, white, nails without splitting, not very strong, straight grained.
Yellow Pine	Virginia to Texas.	Most important lumber for heavy construction and exterior work, keelsons, risings, filling pieces, clamps, floors, bulkheads of small boats, shores, wedges, plugs, strongbacks, staging, joists, posts, piling, ties, paving blocks.	Hard, strong, heartwood is durable in the ground, grain varies, heavy, tough, reddish brown in color, resinous, medullary rays well marked.

with doubling up sometimes used for greatest strength and durability.

Nails (Figs. 1-6 and 1-7) are probably the most primitive and merely the simplest of the current methods.

Driven fasteners would be better, but nails and screws are similar in many respects. Adhesives and glues are also similar enough to be interchangeable words in many contexts.

The third method, which is really the oldest, is simple wood-to-wood joinery—on through some not-so simple ancient and modern joints. See Figs. 1-8 through 1-15 and Tables 1-1 through 1-4.

Table 1-2. Plywood Grades. (Continues on page 16.)

Type	Grade Desingnation[2]	Description and Most Common Use	Typical Grade-trademarks	Face	Back	Inner Plies	Most Common Thicknesses (inch)[3]					
Interior Type	N-N, N-A, N-B INT-APA	Cabinet quality. For natural finish furniture, cabinet doors, built-ins, etc. Special order items.	NN G1 NT APA PS 1 74 / NA G2 INT APA PS 1 74	N	N.A. or B	C						3/4
	N-D-INT-APA	For natural finish paneling. Special order items.	ND G3 INT APA PS 1 74	N	D	D	1/4					
	A-A INT-APA	For applications with both sides on view. Built-ins, cabinets, furniture and partitions. Smooth face; suitable for painting.	AA G4 INT APA PS 1 74	A	A	D	1/4		3/8	1/2	5/8	3/4
	A-B INT-APA	Use where appearance of one side is less important but two smooth solid surfaces are necessary.	AB G4 INT APA PS 1 74	A	B	D	1/4		3/8	1/2	5/8	3/4
	A-D INT-APA	Use where appearance of only one side is important. Paneling, built-ins, shelving, partitions, and flow racks.	A-D GROUP 1 INTERIOR PS 1-74 000 (APA)	A	D	D	1/4		3/8	1/2	5/8	3/4
	B-B INT-APA	Utility panel with two smooth sides. Permits circular plugs.	BB G1 INT APA PS 1 74	B	B	D	1/4		3/8	1/2	5/8	3/4
	B-D INT-APA	Utility panel with one smooth side. Good for backing, sides of built-ins. Industry: shelving, slip sheets, separator boards and bins.	B-D GROUP 3 INTERIOR PS 1-74 000 (APA)	B	D	D	1/4		3/8	1/2	5/8	3/4
	DECORATIVE PANELS-APA	Rough-sawn, brushed, grooved, or striated faces. For paneling, interior accent walls, built-ins, counter facing, displays, and exhibits.	PS 1 74 DECORATIVE BD G1 INT APA	C or btr	D	D		5/16	3/8	1/2	5/8	
	PLYRON INT-APA	Hardboard face on both sides. For counter tops, shelving, cabinet doors, flooring. Faces tempered, untempered, smooth, or screened.	PLYRON INT APA PS 1 74			C & D				1/2	5/8	3/4
Exterior Type	A-A EXT-APA	Use where appearance of both sides is important. Fences, built-ins, signs, boats, cabinets, commercial refrigerators, shipping containers, tote boxes, tanks, and ducts. (4)	AA G3 EXT APA PS 1 74	A	A	C	1/4		3/8	1/2	5/8	3/4
	A-B EXT-APA	Use where the appearance of one side is less important. (4)	AB G1 EXT APA PS 1 74	A	B	C	1/4		3/8	1/2	5/8	3/4
	A-C EXT-APA	Use where the apperance of only one side is important. Soffits, fences, structural uses, boxcar and truck lining, farm buildings. Tanks, trays, commercial refrigerators. (4)	A-C GROUP 3 EXTERIOR PS 1-74 000 (APA)	A	C	C	1/4		3/8	1/2	5/8	3/4
	B-B EXT-APA	Utility panel with solid faces. (4)	BB G1 EXT APA PA 1 74	B	B	C	1/4		3/8	1/2	5/8	3/4
	B-C EXT-APA	Utility panel for farm service and work buildings, boxcar and truck lining, containers, tanks, agricultural equipment. Also as base for exterior coatings for walls, roofs. (4)	B-C GROUP 2 EXTERIOR PS 1-74 000 (APA)	B	C	C	1/4		3/8	1/2	5/8	3/4
	HDO EXT-APA	High Density Overlay plywood. Has a hard, semi-opaque resin-fiber overlay both faces. Abrasion resistant. For concrete forms, cabinets, counter tops, signs and tanks. (4)	HDO 6060 BB PLYRORM EXT	A or B	A or B	C or C plgd			3/8	1/2	5/8	3/4

MDO EXT-APA	Medium Density Overlay with smooth, opaque, resin-fiber overlay one or both panel faces. Highly recommended for siding and other outdoor applications, built-ins, signs, and displays ideal base for paint. (4)	MDO BB G4 EXT APA PS 1 74	B	B or C	C			3/8	1/2	5/8	3/4
303 SIDING EXT-APA	Proprietary plywood products for exterior siding, fencing, etc. Special surface treatment such as V-groove channel groove, striated, brushed, rough-sawn. (6)	303 SIDING 16 oc GROUP 1 EXTERIOR PS 1-74 (APA) 000	(5)	C	C			3/8	1/2	5/8	
T 1-11 EXT-APA	Special 303 panel having grooves 1/4" deep, 3/8" wide, spaced 4" or 8" o.c. Other spacing optional. Edges shiplapped. Available unsanded, textured, and MDO. (6)	303 Siding 16 oc T111 GROUP 1 EXTERIOR PS 1-74 (APA) 000	C or btr.	C						5/8	
PLYRON EXT-APA	Hardboard faces both sides, tempered, smooth or screened.	PLYRON EXT APA PS 1 74			C				1/2	5/8	3/4
MARINE EXT-APA	Ideal for boat hulls. Made only with Douglas fir or western larch. Special solid jointed core construction. Subject to special limitations on core gaps and number of face repairs. Also available with HDO or MDO faces.	MARINE AA EXT APA PS 1 74	A or B	A or B	B	1/4		3/8	1/2	5/8	3/4

Table 1-3. Plywood Types, a Typical Trademark, and Identifiction Indexes for Unsanded Grades.

UA USES, APPLICATIONS			Veneer Grade			Most Common Thicknesses (inch) (2) (3)				
Use These Terms When You Specify Plywood	Description and Most Common Uses	Typical Grade-trademarks	Face	Back	Inner Plies					
C-D INT-APA (1) (4)	For wall and roof sheathing, subflooring, industrial uses such as pallets. Usually supplied with exterior glue; sometimes with interior glue. Product Standard has provision for intermediate glue, but availability is currently limited. Specify exterior glue for best durability in longer construction delays and for treated wood foundations.	C-D 48/24 INTERIOR PS 1-74 000 (APA)	C	D	D	5/16	3/8	1/2	5/8	3/4
STRUCTURAL-I C-D INT APA (4) and STRUCTURAL II C-D INT-APA (4)	Unsanded structural grades where plywood strength properties are of maximum importance: structural diaphragms, box beams, gusset plates, stressed-skin panels, containers, pallet bins. Made only with exterior glue. See (5) for Group requirements.	STRUCTURAL I C-D 32/16 INTERIOR PS 1-74 000 (APA)	C	D	D	5/16	3/8	1/2	5/8	3/4
C-D PLUGGED INT-APA (1) (5)	For built-ins, wall and ceiling tile backing, cable reels, walkways, separator boards. Not a substitute for Underlayment, as it lacks Underlayment's punch-through resistance. Touch-sanded.	C-D PLUGGED GROUP 2 (APA) INTERIOR PS 1-74 000	C Plgd.	D	D	5/16	3/8	1/2	5/8	3/4
2·4·1 INT-APA (1) (6)	"Heavy Timber" roof decking and combination subfloor-underlayment. Quality base for resilient floor coverings, carpeting, wood strip flooring. Use 2·4·1 with exterior glue in areas subject to moisture. Unsanded or touch-sanded as specified.	2·4·1 GROUP 1 (APA) INTERIOR PS 1-74 000	C Plgd.	D	C & D	1-1/8"				
C-C EXT-APA (4)	Unsanded grade with waterproof bond for subflooring and roof decking, siding on service and farm buildings, wood foundations, crating, pallet bins, cable reels.	C-C 42/20 EXTERIOR PS 1-74 000 (APA)	C	C	C	5/16	3/8	1/2	5/8	3/4

STRUCTURAL I C-C EXT-APA (4) and STRUCTURAL II C-C EXT-APA (4)	For engineered applications in construction and industry where full Exterior type panels are required. Unsanded. See (5) for Group requirements.	STRUCTURAL I C-C 32/16 EXTERIOR PS 1-74 000 APA	C	C	C		5/16	3/8	1/2	5/8	3/4

(1) Can be manufactured with exterior or intermediate glue (check dealer for availability of intermediate glue in your area).
(2) All grades can be manufactured tongue-and-grooved in panels 1/2" and thicker.

(3) Panels are standard 4×8-foot size. Other sizes available.
(4) Grade-trademarked with Identification Index shown in guide below

(5) Can be manufactured in Structural I (all plies limited to Group 1 species) and Structural II (all piles limited to Group 1, 2, or 3 species).
(6) Available in Group 1, 2, or 3 only

Typical grade-trademark

Grade of veneer on panel face
Grade of veneer on panel back

Identification Index — C-D
Designates the type of plywood — 32/16
Exterior or Interior — INTERIOR   APA
Product Standard governing — PS 1-74
manufacture — 000
Type of glue used — EXTERIOR GLUE

Mill number

Courtesy of the American Plywood Association.

Identification indexes for unsanded grades
Panels thicker than 7/8 inch shall be identified by Group.

Thickness (inch)	C-C EXTERIOR C-D INTERIOR			STRUCTURAL I C-D & C-C STRUCTURAL II(b) C-D & C-C	STRUCTURAL II C-D & C-C
	Group 1 Group 2(a)	Group 2 or 3 Group 4(a)	Group 4	Group 1	Group 2 or 3
5/16	20.0	16.0	12.0	20.0	16.0
3/8	24.0	20.0	16.0	24.0	20.0
1/2	32/16	24.0	24.0	32/16	24.0
5/8	42/20	32/16	30.12	42/20	32/16
3/4	48/24	42/20	36/16	48/24	42/20
7/8		48/24	42.20		38/24

(a) Panels conforming to special thickness provisions and panel construction of Paragraph 3.8.6 of PS 1.
(b) Panels manufactured with Group 1 faces but classified as STRUCTURAL II by reason of Group 2 inner plies.

(1) Sanded both sides except where decorative or other surfaces specified.
(2) Available in Group 1, 2, 3, 4, or 5 unless otherwise noted.
(3) Standard 4×8 panel sizes, other sizes available.
(4) Also available in Structural I (all plies limited to Group 1 species) and Structural II (all plies limited to Group 1, 2, or 3 species).
(5) C or better for 5 plies; C Plugged or better for 3-ply panels.

(6) Stud spacing is shown on grade stamp.
(7) For finishing recommendations, see form V307.
(8) For strength properties of appearance grades, refer to "Plywood Design Specification," form Y510.

Courtesy of the American Plywood Association

Table 1-4. Wood Species Used in Plywoods.

Group 1	Group 2		Group 3	Group 4	Group 5
Apitong (a)(b)	Cedar, Port Orford	Maple, Black	Alder, Red	Aspen	Basswood
Beech, American	Cypress	Mengkulang (a)	Birch, Paper	Bigtooth	Fir, Balsam
Birch	Douglas Fir 2 (c)	Meranti, Red (a)(d)	Cedar, Alaska	Quaking	Poplar, Balsam
Sweet	Fir	Mersawa (a)	Fir, Subalpine	Cativo	
Yellow	California Red	Pine	Hemlock, Eastern	Cedar	
Douglas Fir 1 (c)	Grand	Pond	Maple, Bigleaf	Incense	
Kapur (a)	Noble	Red	Pine	Western Red	
Keruing (a) (b)	Pacific Silver	Virginia	Jack	Cottonwood	
Larch, Western	White	Western White	Lodgepole	Eastern	
Maple, Sugar	Hemlock, Western	Spruce	Ponderosa	Black (Western Poplar)	
Pine	Lauan	Red	Spruce	Pine	
Caribbean	Almon	Sitka	Redwood	Eastern White	
Ocote	Bagtikan	Sweetgum	Spruce	Sugar	
Pine, Southern	Mayapis	Tamarack	Black		
Loblolly	Red Lauan	Yellow Poplar	Engelmann		
Longleaf	Tangile		White		
Shortleaf	White Lauan				
Slash					
Tanoak					

(a) Each of these names represents a trade group of woods consisting of a number of closely related species.
(b) Species from the genus Dipterocarpus are marketed collectively. Apitong if originating in the Philippines; Keruing if originating in Malaysia or Indonesia.
(c) Douglas fir from trees grown in the states of Washington, Oregon, California, Idaho, Montana, Wyoming, and the Canadian Provinces of Alberta and British Columbia shall be classed as Douglas fir No. 1. Douglas fir from trees grown in the states of Nevada, Utah, Colorado, Arizona and New Mexico shall be classed as Douglas fir No. 2.
(d) Red Meranti shall be limited to species having a specific gravity of 0.41 or more based on green volume and oven dry weight.

Chapter 2

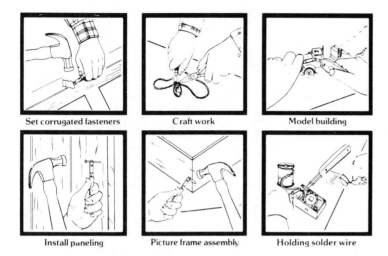

Set corrugated fasteners   Craft work   Model building

Install paneling   Picture frame assembly   Holding solder wire

Driven Fasteners for Wood

There are obvious differences between screws and nails. From that point, it becomes a relatively complex job to select the proper fastener to do the job—at the right cost, in the right time, with the right effort. Nails are basically wedges that become fasteners as they are driven with another tool.

Carpenter's hammers come in ripping-claw and curved-claw models, and widely varying head weights, from about 13 ounces on up to 32 ounces. See Fig. 2-1. Handle styles vary so it pays to feel among different brands and different handle materials to make a wise selection of the tool you will use for general-purpose nailing. My preference is for a 20-ounce, fiberglass-handled model with a straight claw for general work.

The standard head weight of 16 ounces makes a decent starting point for most people. The head is neither too light to drive larger nails nor too heavy to work with on most brads and light material. See Figs. 2-2 through 2-6.

You have a choice of wood, fiberglass, or steel, and in the case of steel you can use a hollow, tubular style or a solid-steel handle. The steel is the strongest, but any will withstand the rigors of daily use by professionals if the hammer is of good quality in the first place. Nail pulling is not the primary job of the carpenter's hammer. I start to get a little edgy about pulling nails with a hammer when I reach the 10d nail size. At that point I find myself looking for a

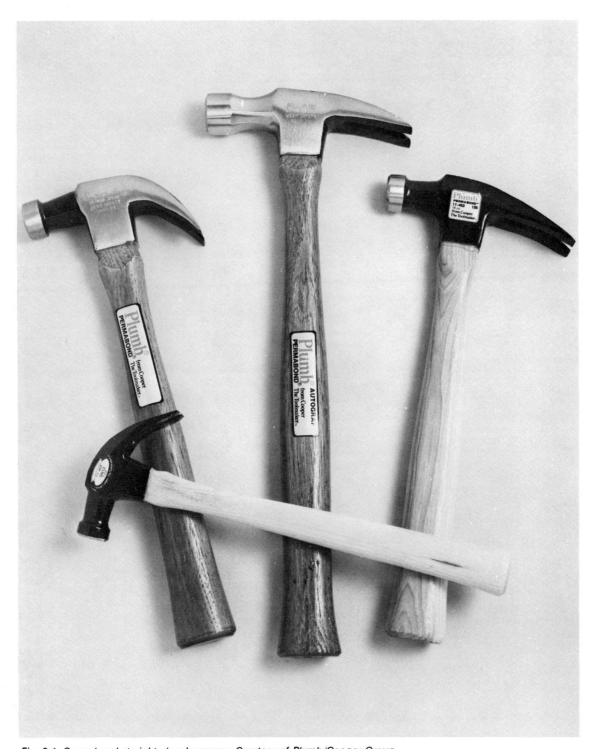

Fig. 2-1. Curved and straight claw hammers. Courtesy of Plumb/Cooper Group.

Fig. 2-2. A 16-ounce curved claw hammer, with a fiberglass handle (top), a rip claw hammer, with a fiberglass handle (left), and a ball peen hammer, with fiberglass handle (right). Courtesy of The Stanley Works.

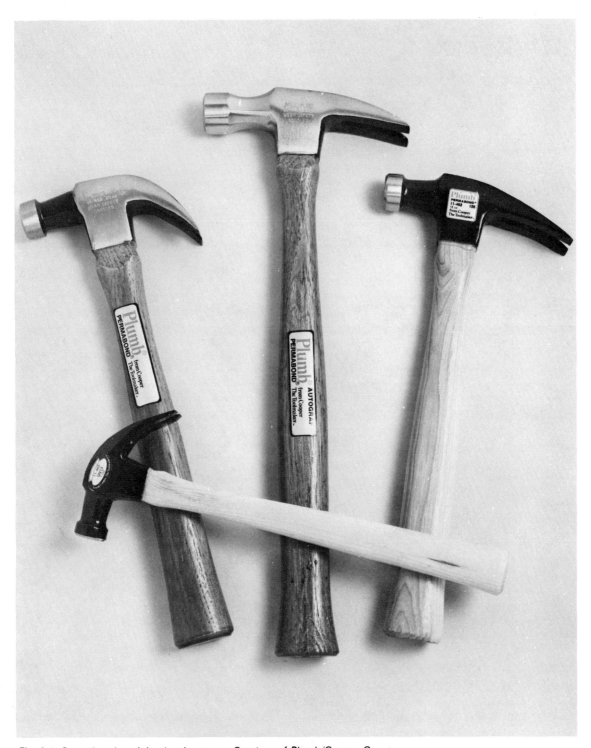

Fig. 2-1. Curved and straight claw hammers. Courtesy of Plumb/Cooper Group.

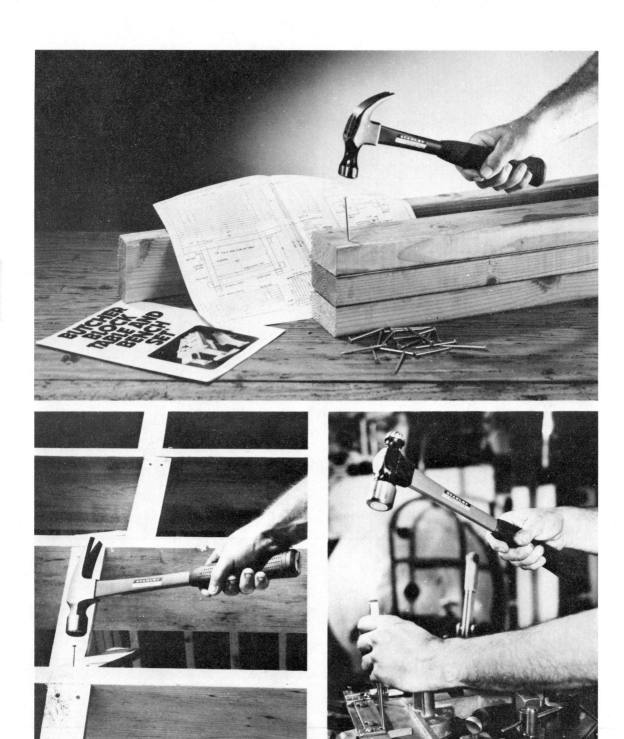

Fig. 2-2. A 16-ounce curved claw hammer, with a fiberglass handle (top), a rip claw hammer, with a fiberglass handle (left), and a ball peen hammer, with fiberglass handle (right). Courtesy of The Stanley Works.

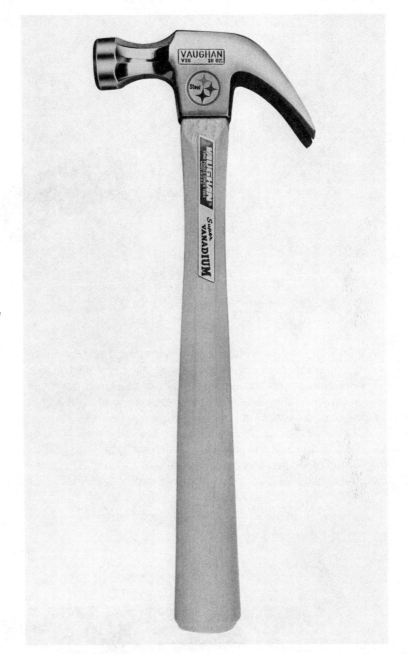

Fig. 2-3. A 16-ounce hickory handle, curved-claw hammer. Courtesy of Vaughan & Bushnell.

nail-puller, crowbar, or other prytool of solid steel. See Fig. 2-7. Much depends on how deeply the nail is driven, whether or not it has a ring shank, and whether or not the thing is in solid wood or just driven on through a facing piece.

All steel and fiberglass handles will have plastic (usually neoprene) or rubber cushion handles to help absorb some of the shock of

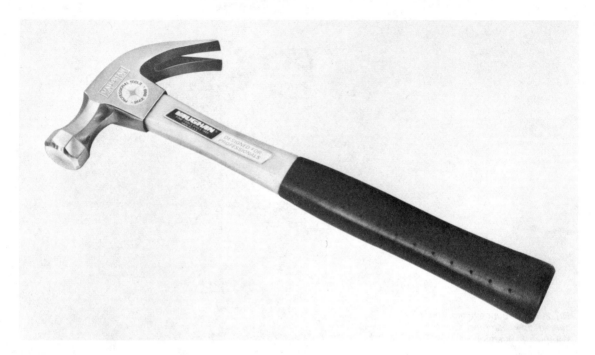

blows. Check these for comfort, but keep in mind you can make minor shape changes in almost any hammer handle. If the hammer suits in all other respects, a slight hand discomfort over a long job can be corrected by trimming the neoprene pad with rough sandpaper or doing the same job to a wood handle.

　　If the hammer feels out of balance when you try to swing it, back off and think. The problem might be your grip, but you also might be more suited to a different brand of hammer. There are different methods of shaping handles, each with a slightly different

Fig. 2-4. A glass-handled, curved-claw hammer. Courtesy of Vaughan & Bushnell.

Fig. 2-5. A tubular-steel-handled, curved-claw hammer. Courtesy of Vaughan & Bushnell.

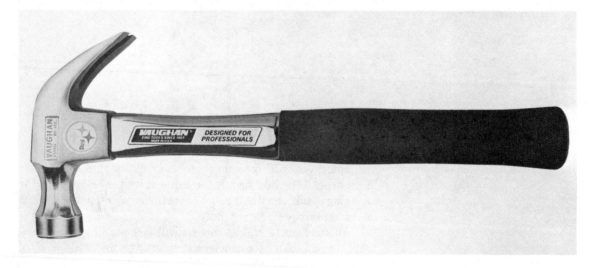

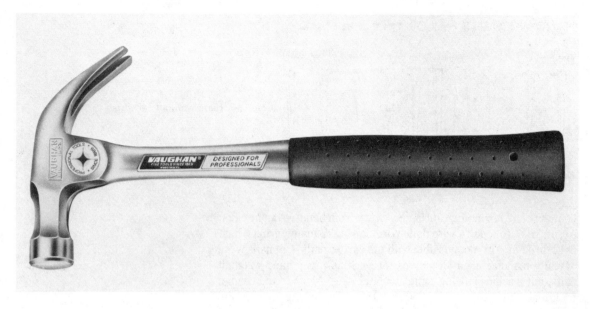

Fig. 2-6. A solid-steel, curved-claw hammer. Courtesy of Vaughan & Bushnell.

"feel." Select the one that suits you best rather than one that friends or ads convince you is best.

Head styles vary from round to octagon on the rim of the striking face. Most have a slight bell to let you drive nails flush without damaging the surrounding material. Cross-hatched or mill-finished faces are supposed to keep nails from flying quite as much. I've got several of these hammers and I generally use them on rough framing work. Don't use them where you don't want hammer marks to show.

Heads join handles in various ways. Most wood handles are held in by wedges and by epoxy to add strength to the bond. Fiberglass and some steel handles will be epoxy bonded. Just check the fit and finish carefully. In most all cases with a well-made hammer, the handle will break long before it loosens at the head.

Look for detailing marks of general finish that indicate care paid during making. Examples are carefully ground tips on the claws and a carefully made notch where the claws join. Polishing, if any, should be well done over the area it covers, and paint should cover evenly when used. Look for no bubbles at all in the epoxy around the eye; occasionally a few smaller ones do slip by on well-made tools.

NAILS

Fig. 2-7. A cat's claw nail puller. Courtesy of The Stanley Works.

Old hand-wrought nails (Figs. 2-8 through 2-11) are still found in

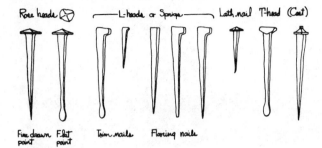

Rose heads ⊕ ——— L-heads or Sprigs ——— Lath nail T-head (Cut)

Fine drawn Flat Trim nails Flooring nails
point point

Fig. 2-8. Hand-wrought nail styles.

country. The technological changes were from hand-wrought nails, to cut nails, to today's wire nail. Wire nails take many forms (Figs. 2-12 through 2-18) not possible with the earlier nails. Cut nails were actually manufactured already as far back as many hand-wrought nails, but machine cutting came at a later date. Much of the change took place between about 1790 and 1830. The changeover in the United States seems to have taken place slightly ahead of that of England, but U.S. Patent Office dating on old records was destroyed in a fire in 1836.

Early machine cut nails had handmade heads. By the latter part of the changeover period, the entire nail was made by machine. Modern machinery makes nails in much the same way as machinery from the 1830s. Tremont Nail Co. (P.O. Box 111, Wareham, MA

Fig. 2-9. Heading hand wrought nails.

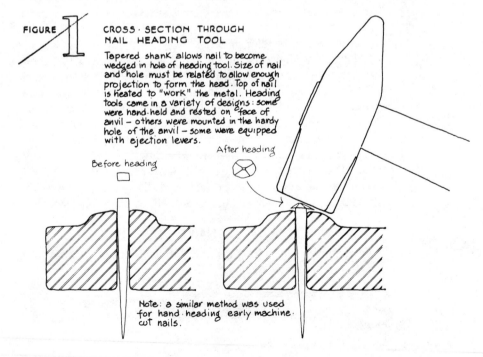

FIGURE 1

CROSS · SECTION THROUGH NAIL HEADING TOOL

Tapered shank allows nail to become wedged in hole of heading tool. Size of nail and hole must be related to allow enough projection to form the head. Top of nail is heated to "work" the metal. Heading tools came in a variety of designs: some were hand-held and rested on face of anvil — others were mounted in the hardy hole of the anvil — some were equipped with ejection levers.

Before heading

After heading

Note: a similar method was used for hand-heading early machine-cut nails.

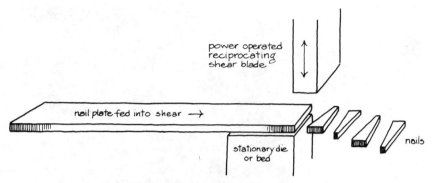

power operated
reciprocating
shear blade

nail plate fed into shear →

stationary die
or bed

nails

SIMPLIFIED DIAGRAM SHOWING THE BASIC
PROCESS OF MAKING EARLY CUT NAILS

Fig. 2-10. Making early cut nails.

02571) is one of the major manufacturers of cut nails today. The company was started in 1819. Today, about 50 machines are producing a wide variety of cut nails for both general building and decorative purposes, including restorations.

Today's cut nails come in wide varieties, and they offer some distinct advantages over wire nails. According to Tremont Nail Company, cut nails have much more holding power than wire nails. The wedge shape and rougher edges all help here, and in some cases the added holding power can be 70 percent more than a similar size wire nail. Bending is a little less of a problem even with untempered cut nails and almost no problem at all with tempered cut nails. Sometimes in masonry work even tempered nails will bend. Remember, too, that tempered nails are to be driven with a hand-drilling hammer or an engineer's or blacksmith's hammer—not a claw hammer. Cut nails are often clinched, when untempered, on the opposite side of boards. They tend to look somewhat more historically attractive in such uses than do wire nails.

The variety of cut nails is wide, and quoting just from the list made by Tremont provides a good example. Hillwood Manufacturing makes some similar nails, but does not go into the restoration and historical types that Tremont Nail Company makes.

Spikes are used for heavy construction involving planks and heavy timbers, often used with small bridges, and with post and beam framing styles for residences.

Masonry nails are used to nail furring strips and other materials to cement block, mortar joints, new concrete, and brick walls. You'll find that with old concrete work where the material has had years to cure, the concrete is hard enough to almost make essential the use of a tool such as the Remington Power gun. This tool uses a .22-caliber blank cartridge to drive a steel pin of appropriate length. It is a far better way of working on masonry, but it's far more expensive.

Fig. 2-11. Nail mill operations. Courtesy of Tremont Nail Co.

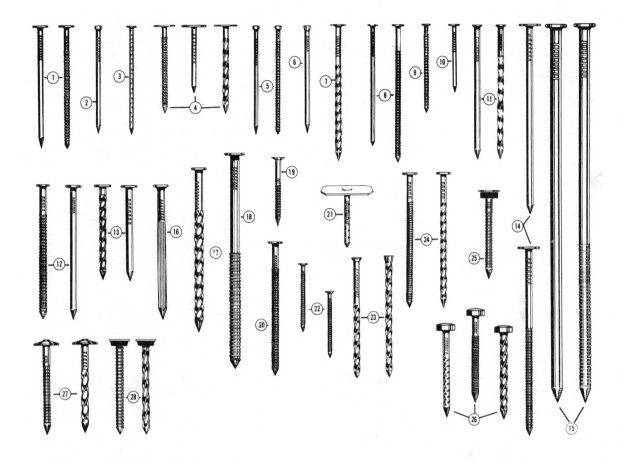

STORMGUARD® NAILS
FOR EXTERIOR APPLICATIONS
(Hot-dipped zinc-coated twice in molten zinc)

1. Wood Siding, Box
   (Plain & Anchor)
2. Finishing
3. Insulating, Plastic Siding
4. Asphalt Shingle
   (Anchor, Plain & Screw)
5. Cedar Shake (Plain & Anchor)
6. Casing
7. Cribber
8. "Split-Less" Wood Siding
   (Plain & Anchor)
9. Asbestos
10. Cedar Shingle
11. Hardboard Siding
    (Plain & Screw)
12. Common (Anchor & Plain)
13. Aluminum, Steel & Vinyl Siding
    (Screw & Plain)
14. Insulation Roof Deck
    (Plain & Anchor)
15. Gutter Spike (Plain & Anchor)

INTERIOR & OTHER NAILS

16. Masonry
17. Post-Barn/Truss Rafter
    (Screw)
18. Post-Barn (Anchor)
19. Drywall, GWB-54 Style
20. Underlayment, Plywood
    (Sub-floor, sheathing, etc.)
21. "Square-Cap" Roofing
22. Underlayment
    (Flat Head & Countersunk)
23. Spiral Flooring
    (Casing Head & Countersunk)
24. Pallet (Anchor & Screw)

METAL ROOFING NAILS

25. Rubber Washer
    (Stormguard, Anchor)
26. Compressed Lead Head
    (Barbed, Anchor & Screw)
27. Umbrella Head
    (Stormguard, Anchor & Screw)
28. Lead Washer
    (Stormguard, Anchor & Screw)

Penny-Wise Nail Lengths

2d	1″	12d	3¼″
3d	1¼″	16d	3½″
4d	1½″	20d	4″
5d	1¾″	30d	4½″
6d	2″	40d	5″
7d	2¼″	50d	5½″
8d	2½″	60d	6″
9d	2¾″	70d	7″
10d	3″	80d	8″

Fig. 2-12. Cut nail types for today. Courtesy of Tremont Nail Co.

LENGTHS AND GAUGE OF SMOOTH SHANK WIRE NAILS BY "PENNY" DESIGNATION

Penny Size	COMMON NAILS*			SMOOTH BOX NAILS*			SINKERS*			COOLERS*		
	Length	Gauge	Approx. No. Per Lb.	Length	Gauge	Approx. No. Per Lb.	Length	Gauge	Approx. No. Per Lb.	Length	Gauge	Approx. No. Per Lb.
2d	1	15	847	1	15-1/2	940	–	–	–	–	–	–
3d	1-1/4	14	543	1-1/4	14-1/2	588	1-1/8	15-1/2	923	1-1/8	16	978
4d	1-1/2	12-1/2	294	1-1/2	14	453	1-3/8	14	527	1-3/8	15-1/2	680
5d	1-3/4	12-1/2	254	1-3/4	14	389	1-5/8	13-1/2	387	1-5/8	15	570
6d	2	11-1/2	167	2	12-1/2	225	1-7/8	13	293	1-7/8	13-1/2	315
7d	2-1/4	11-1/2	150	2-1/4	12-1/2	200	2-1/8	12-1/2	223	2-1/8	13-1/2	280
8d	2-1/2	10-1/4	101	2-1/2	11-1/2	136	2-3/8	11-1/2	153	2-3/8	12-1/2	190
9d	2-3/4	10-1/4	92	2-3/4	11-1/2	124	–	–	–	2-7/8	11-1/2	119
10d	3	9	66	3	10-1/2	90	2-7/8	11	111	–	–	–
12d	3-1/4	9	61	3-1/4	10-1/2	83	3-1/8	10	81	–	–	–
16d	3-1/2	8	47	3-1/2	10	69	3-1/4	9	64	–	–	–
20d	4	6	29	4	9	50	3-3/4	7	39	–	–	–
30d	4-1/2	5	22	4-1/2	9	44	4-1/4	6	29	–	–	–
40d	5	4	17	5	8	33	4-3/4	5	22	–	–	–
50d	5-1/2	3	13	–	–	–	–	–	–	–	–	–
60d	6	2	10	–	–	–	5-3/4	3	13	–	–	–

*Above items are listed for comparison purposes only and are not produced by the Hillwood Mfg. Co., **BUT ARE AVAILABLE FROM HILLWOOD.**

MEASUREMENT INFORMATION — LENGTHS AND GAUGES

LENGTH

Flat Head and Oval Head Drive Screws and Nails are measured from under the head and include the point to determine the length of the shank.

Countersunk Head and Casing Head Drive Screws and Nails are measured from the top of the head to the end of the point to determine the length.

GAUGES

The Screw Gauges, Wire Gauges and Decimal Gauges referred to on the Helyx (Tempered) Drive Screws. Drive Screws are measured on the outside diameter of the threads on the shank.

The wire gauges and decimal gauges referred to in the Rol-Thread and Ring Barb Nails are measured on the smooth shank under the head and designate the size wire from which they are produced and not the outside diameter of the threads on the shank.

Common nails are used for framing, roughing-in for plumbing and wiring, and for face nailing floors and other materials.

Headless *foundry* nails are used as metal wedges, for building molds, and on other chores in metal-casting work.

Boat nails are always hot-dipped galvanized for use in building wooden boats. Corrosion resistance is great and so is holding power.

Sheathing nails resemble common nails quite closely, but the

Fig. 2-13. *Nail penny sizes. Courtesy of Hillwood Mfg. Co.*

Fig. 2-14. *Numbers of nails per pound. Courtesy of Hillwood Mfg. Co.*

Steel Wire Gauge	3/16	1/4	3/8	1/2	5/8	3/4	7/8	1	1-1/8	1-1/4	1-1/2	1-3/4	2	2-1/4	2-1/2	2-3/4	3	3-1/2	4	4-1/2	5	6
2								60	54	48	41	35	31	28	25	23	21	18	16	14	13	11
3								67	60	55	47	41	36	32	29	27	25	21	18	16	15	12
4								81	74	66	55	48	41	37	34	31	29	25	22	20	18	15
5								90	81	74	61	52	45	41	38	35	32	28	24	22	21	18
6				213	174	149	128	113	101	91	76	65	58	52	47	43	39	34	29	26	24	20
7				250	205	174	148	132	120	110	92	78	70	61	55	53	51	40	35	31	28	24
8				272	238	198	174	153	139	126	106	93	82	74	66	61	56	48	42	38	34	28
9					348	286	238	213	185	152	128	112	99	87	79	71	67	58	50	45	44	34
10				469	373	320	277	242	216	196	165	142	124	111	100	91	84	71	62	55	49	42
11				510	417	366	323	285	254	233	200	171	149	136	122	111	103	87	77	69	61	52
12				740	603	511	442	405	351	327	268	229	204	182	161	149	137	118	103	95	87	71
13			1356	1017	802	688	508	458	412	348	294	260	232	209	190	175	153	138	123	110	93	
14		2293	1664	1290	1037	863	806	667	610	536	459	406	350	312	278	256	233	201	176	157	140	117
15		2890	2213	1619	1316	1132	971	869	787	694	578	501	437	390	351	317	290	246	220	196	177	145
16		3932	2720	2142	1708	1414	1229	1090	973	872	739	635	553	496	452	410	370	318	277	248	226	
17		5316	3890	2700	2306	1904	1581	1409	1253	1139	956	831	746	666	590	532	486	418	360	322	295	
18		7520	5072	3824	3130	2608	2248	1976	1760	1590	1338	1150	996	890	820	740	680	585	507	448	412	
19		9920	6860	5075	4132	3508	2816	2556	2284	2096	1772	1590	1390	1205	1060	970	895	800				
20	18620	14050	9432	7164	5686	4795	4230	3596	3225	2893	2412	2070	1810	1620	1450	1315	1215	1435				

Applications	Common Nails					Threaded Nails & Drive Screws			
	Size D	Length In.	Diam., In.	Number or Spacing	No. per Lb.	Length In.	Diam., In.	Number or Spacing	No. per Lb.
Joists to sill or girder, toe nail	16	3-1/2	.162	3	49	3-1/4	.135	3	73
1 x 6 subfloor to joists, face nail	8	2-1/2	.131	2	101	2-1/8	.109	2	171
1 x 8 subfloor to joists, face nail	8	2-1/2	.131	3	101	2-1/8	.109	2	171
Sole plate to joists or header joist	16	3-1/2	.162	16" OC	49	3-1/4	.135	16" OC	73
Stud to sole plate, toe nail	16	3-1/2	.162	3	49	2-1/2	.120	3	117
Stud to sole plate, toe nail	10	3	.148	3	69	2-1/2	.120	3	117
Stud to sole plate, end nail	16	3-1/2	.162	2	49	3-1/4	.135	2	73
Top plate to studs, end nail	16	3-1/2	.162	2	49	3-1/4	.135	2	73
Doubled studs	16	3-1/2	.162	24" OC	49	2-1/2	.120	16" OC	117
Doubled studs	16	3-1/2	.162	30" OC	49	2-1/2	.120	16" OC	117
Top plates, spike together	16	3-1/3	.162	24" OC	49	3-1/4	.135	24" OC	73
Ceiling joists to plate, toe nail	16	3-1/2	.162	2	49	3-1/4	.135	2	73
Rafter to plate	16	3-1/2	.162	3	49	3-1/4	.135	3	73
1" x 8" sheathing	8	2-1/2	.131	2	101	2	.120	2	147
Plywood:									
Subfloor to joists, edges	8	2-1/2	.131	6" OC	101	2-1/8	.109	6" OC	171
Subfloor to joists, intermediate	8	2-1/2	.131	10" OC	101	2-1/8	.109	12" OC	171
Roof sheathing, to rafters, edges	6	2	.120	6" OC	181	1-3/4	.120	6" OC	165
Roof sheathing, to rafters, intermediate	6	2	.120	12" OC	181	1-3/4	.120	12" OC	165
Horizontal bevel siding	8	2-1/2	.131		101	2-1/8	.115		182

Fig. 2-15. Nailing applications.

shank is thinner. They're used to attach all kinds of sheathing to wood studs and for face-nailing thick board flooring.

Rose-head clinch nails are made with a button head. They are left malleable enough to make clinching easy when it is desirable. Such clinching is most often used when you're building board doors,

Length	Pkg.*	20	19	18	17	16	15	14	13	12
3/8	50	X	X	X						
	25	X	X	X						
	1	X	N/A	X						
1/2	50	X	X	X	X	X				
	25	X	X	X	X	X				
	1	X	X	X	X	X				
5/8	50	X	X	X	X	X				
	25	X	X	X	X	X				
	1	X	X	X	X	X				
3/4	50	X	X	X	X	X	X	X		
	25	X	X	X	X	X	X	X		
	1	X	X	X	X	X	X	X		
7/8	50	X	X	X	X	X	X			
	25	X	X	X	X	X	X			
	1	X	X	X	X	X	X	N/A		
1	50	X	X	X	X	X	X	X		
	25	X	X	X	X	X	X	X		
	1	X	X	X	X	X	X	X		
1-1/4	50			X	X	X	X	X		
	25			X	X	X	X	X		
	1			X	X	X	X	X		
1-1/2	50			X	X	X	X	X		
	25			X	X	X	X	X		
	1			X	X	X	X	X		
1-3/4	50					X	X	X	X	X
	25					X	X	X	X	X
	1					X	X	X	N/A	N/A
2	50					X	X	X	X	X
	25					X	X	X	X	X
	1					X	X	X	X	X
STEEL WIRE BRADS										
2-1/4	50						X	X	X	X
	25						X	X	X	X
	1						X	X	N/A	N/A
2-1/2	50						X	X	X	X
	25						X	X	X	X
	1						X	X	X	X

*50 lb. Bulk — 25 lb. Bulk — 1 lb. Boxes
N/A — Not Available

Fig. 2-16. Brads. Courtesy of Hillwood Mfg. Co.

The Maze Stormguard® Way . . .

Nails completely immersed into a vat of molten zinc (entire nail gets coated).

1) Hot-Dipping in Molten Zinc is acknowledged by most authorities as the **best way** to apply a heavy uniform coating to nails. In this process the nails are completely immersed into a vat of molten zinc, similar to french-frying potatoes. This not only gives an outer coating of pure zinc, but also provides a tenacious inner coating of zinc-steel alloy. Our Stormguard® nails are further protected by a second dip into the protective zinc to preclude pinholes or imperfections — and to increase the amount of zinc on every nail. This exclusive processing by completely automated equipment insures both a heavy, uniform zinc coating and clean threads for high holding power.

The proof is the record of Maze hot-dipped nails on thousands and thousands of structures built over the last **60 years** — not a single verified report of stains or streaks due to rust. No wonder contractors say that Stormguards are the best steel-base nails on the market — economical, easy driving, and specially designed to properly apply all types of modern roofing, siding and trim.

Rotating nails with zinc flakes in a barrel in a furnace.

2) Hot-Galvanized (Tumbler or Barrel Process) nails are the type most often produced by big steel mills. Unfortunately, the term "hot-galvanized" is confused by many people with "hot-dipped" nails. But "hot-galvanized" nails are **not** dipped and do not offer the same performance as do hot-dipped nails. Hot-galvanized nails are coated by sprinkling zinc chips on cold steel nails in a barrel, then rotating the hot barrel in a furnace to melt and distribute the zinc. The melting zinc washes off on the nails — somewhat **like buttering popcorn** — and in the same way some nails receive excess zinc and some very little. One difficulty is that the threads of ring and screw shank nails tend to fill up with zinc during this process. Also, it is difficult to deposit enough zinc on top of and under the nail heads. Because this process simply does not produce a uniform quality of coating, it has been abandoned entirely at the Maze plant. We make only hot-dip nails — "double-dipped" Stormguard nails.

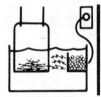

Cold nails rolled in a barrel with zinc dust, glass BB's and activator fluid.

3) Mechanical plated nails, also called **peen-plated** and **golden-galvanized** nails, are a relatively new development involving rolling cold nails around in a barrel with zinc dust, tiny glass "BB's" and an activator fluid. The barrel rotates and the BB's hammer or peen the zinc dust onto the nails. Again the result is relatively clean threads but normally only a **thin deposit** of zinc which is further buttressed by immersion in a chromate rinse which gives a gold or greenish color to the nails. Maze engineers evaluated this process with the inventors until their own tests showed that our exclusive Stormguard double-dipping process was superior for coating outdoor nails. So we do no mechanical plating. We "double-dip" our Stormguard nails.

Basket of nails immersed into an electrolytic solution.

4) Electro-plating of nails is performed by immersing them in a basket into an electrolytic solution so that a **thin film** of zinc is deposited by an electrical current from zinc anodes onto the surface of the nails. Although electroplated nails are beautiful and shiny, it is not feasible economically to build up a heavy enough coating to make this type nail dependably corrosion-resistant for outdoor use. The thin zinc coating soon oxidizes away so that electro-plated nails normally **rust quickly** upon exposure to the weather . . . just ask any experienced siding applicator. In short, plated nails have good holding power from clean threads, but they do not have the heavy zinc coating needed to avoid rusting and staining. We "double-dip" our Stormguard nails

Maze recommends the process proven by so many years of actual field exposure to be the best way to put protective zinc on a strong steel base — **hot-dipping!** We market our hot-dipped line under the trademark, Maze STORMGUARD* — the nails we guarantee to be **hot-dipped twice in molten zinc** by our exclusive process. If you haven't tried them, this is a good time to start.

but these nails are also used for furniture, cabinet work, interior and exterior siding, face nailing wide board flooring, paneling, and in wood fences, with the head left exposed as part of the decor.

Flooring nails are for laying tongue and groove hardwood floors and softwood flooring. Hardened steel is used for hardwood floors and standard nails are used for softwood.

Siding nails look very much like sheathing and common nails,

Fig. 2-17. Nail coating methods. Courtesy of W.H. Maze Co.

Fig. 2-18. Right: Wear safety glasses when working with all striking and struck tools. Courtesy of Vaughan & Busnell.

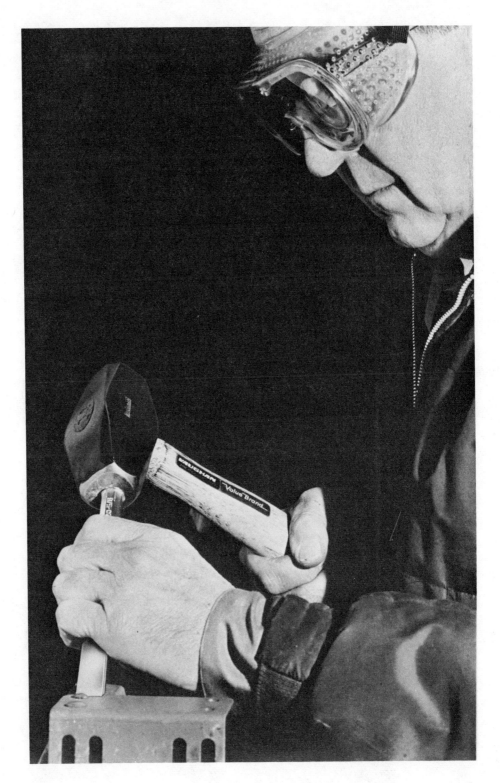

but the shank is thinner and they are normally galvanized to prevent rust streaks and rust through.

Finish nails in cut-nail styles are the same as unhardened flooring nails, and they are also used as casing nails. Finish nails are available in hot-dipped galvanized coating for outdoor work.

Clout nails are similar to roofing nails, but are made of lighter metal. They can be used for shingle applications, box and crate building, and for installing thin siding and paneling. They are also good for some types of furniture repair, cabinet work, making batten doors, and similar building needs. Hot-dipped galvanized nails are stocked by the makers.

Box nails are lighter in construction than common nails but otherwise similar in design. Their original use for making boxes of wood has pretty much gone by the boards these days. Primarily they are used with fences, siding, face nailing of floors, and framing work.

Hinge nails are specifically designed for installing old-fashioned hinges. With a head style similar to that of the boat nail, they are made from lighter-gauge steel.

Slating nails are hot-dipped galvanized finish nails used to repair or install slate roofing.

Fire door clinch nails are designed for the construction of laminated fire doors, clapboard siding, and fencing. Hot-dipped galvanized finish is recommended and stocked for outside use.

Shingle nails are hot-dipped galvanized, and are specifically designed for installing wood shingles and shakes.

Common rose-head nails are used for all the jobs standard common nails are used for, but the decorative head shows up far more nicely for interior and exterior work when the nailhead must be exposed or when exposure is desired as part of a decorative scheme. Floor face nailing is especially attractive with these heads, and the material is available in hot-dipped galvanized finish.

Brads similar in appearance to common and other lighter-weight nails are good for cabinet and furniture applications.

Fine finish nails are useful in construction where a slender, brad-headed nail of greater length than a brad is required. They do well on window casings, door casings, and so on.

Wrought-head nails are designed to look like the hand-forged nail of the late 1700s, with a three-sided head. Tremont Nail Company makes these available in galvanized, brass-plated, or black oxide finishes. They're useful for rough-sawn interior and exterior work with exposed nailheads, face nailing of flooring, and other work where the heads will be exposed as part of the decorative idea for a house or other building. Brass-plated, wrought-head nails are *not* recommended for outdoor use, and I would not recommend them for flooring use.

Tremont Nail Company is the only source I now know of for copper nails and copper decorative tacks. They also distribute copper boat nails and roves from Norway. These solid-copper boat nails and roves offer a square shape. A dished rove (copper boat nails are driven in a drilled hole) is inserted over the shank, which is then nipped to length (the rove has to be driven over the shank with the nailhead supported), and clinched over the rove. They're expensive, but I doubt you'll find many stronger ways to nail (or screw) a boat together.

Rove sizes run from ⅜ inch to ⅝ inch, with nail length up to 5 inches, and shank thickness of 14 to 6 gauge. They're sold by the kilogram, or about 2.2 pounds, with the smallest nail offering about 700 units per kilo and the largest about 500. Cost in most cases is on a 100 count, *per piece*. Thus, 1400 units would make a kilo of ⅜-inch rove boat nails, at the current price per 100 ($1.75). Tremont will break kilos most of the time, but normal packing is one kilo, though half-kilo is readily available.

Wire nails began as less expensive substitutes for cut nails. Early wire nails differ from today's nails in that the heads tend to be bulbous and off-center in relation to the shank. Early wire nails were not made for construction. These were smaller size nails for cigar boxes and such, and comments on the "newer" type of nail— wire nails versus cut nails—were still being made in articles around 1888. At this time, the wire nails had several claimed advantages (mostly in price), and there were about 13 different varieties listed in sizes from 2d on up to 60d.

Sizing nails is generally done on the penny system. Nevertheless, most manufacturers now list products by actual length and wire diameter in preference to the older system. Penny sizing of nails is an old method of determining how many cents a hundred nails would cost. Not a pound, not a hundredweight, but 100 nails. Inflation being what it is, that cost didn't last long, but the sizing system has hung on for a long, long time. Thus penny sizes, beginning at 2 penny (d is the penny abbreviation) and going to 60d, is still with us. Anything larger than 60d is considered a spike (if you've driven many nails over 30d, you'll consider those to be spikes too, I promise), and listed, most of the time, by length and wire diameter. Usually also included are descriptions of the jobs for which the spike is intended (post and beam construction, pole building construction, etc.). Most nails come in sizes to 8 inches long, and special orders can get you greater length.

A post barn nail, as produced by W.H. Maze, is offered in sizes ranging from 2 inches to 8 inches, either just stiffened or hardened, and in wire gauges from minimum of 11 on down to 5½. Head sizes range from 9/32 of an inch to 27/64 of an inch (both 7-inch and 8-inch nails). Gutter spikes are available in 8-inch, 9-inch, and 10-inch lengths, with nail shank diameters at 1/4 inch for those sizes. Ring or plain shanks are available. The spikes are zinc coated (hot-dipped) and have checkered heads. When you start working with spikes in this size range, pound counts shrink wildly. Common nails in 10d (3-inch) length run some 75 or 80 to the pound, but 10-inch spikes run 6 or 7. Eight-inch spikes will run no more than 14 to the pound, with a fairly narrow shank, and as low as 8 with a thicker shank.

Common nails generally stop at the 60d or 6-inch length, with specialty spikes used after that point. You can find a nail, or brad, to fit almost any job by checking various brands. Price differentials are minimal.

One thing I've definitely found is that scrimping on nail selection to save a buck or two on a job is simply not worthwhile. No matter what the job, there's an appropriate nail and that nail style and size should be used. Otherwise, you end up with lifted boards, split wood, or poor holding power.

It's simply not worth saving, for example, $10 or $15 in the cost of building a deck to go with the wrong nails. They must be hot-dipped galvanized, and also screw or other deformed shank—not concrete dipped and tar coated. If this isn't carried through, you can bet on deck flooring lifting at the edges as it contracts and expands over time, and you can also bet on rust streaks.

Rust streaks are less of a problem if you're using opaque stains or paint on a job, but most decks are not so covered. If the wood is left to age to a natural color, whether you're using redwood or pressure-treated wood, within six months nongalvanized nails will provide rust streaks over almost all vertical surfaces and rust spots at the nailheads on horizontal surfaces such as floor. Aluminum nails will not cause this problem. In time, the nailheads will rust off, or the shanks will rust through where the board joint is and your deck will have structural problems. Because many commercially built decks are designed to provide an absolute minimum of strength to suit uses, any such loss is serious. Home-designed decks tend to be a lot stronger.

Common nails are for general-purpose nailing from framing on through installing some types of flooring. They're available in several shank styles—including round, ring, and screw—and may come stiffened, hardened, galvanized, or with slenderized shanks—and in sizes, generally, from 2 to 6 inches.

Post and *truss* nails are similar to common nails, but most usually come with a spiral (screw) shank. They're designed to prevent a truss or post and beam system from giving in to racking forces. They do quite well at that.

Wood siding nails are generally designed for use in woods that tend to split easily when nailed near edges and ends (even with these nails, it makes sense to dull the nail point with a tap and to drill starter holes to prevent splitting of many woods). The shank is slimmer than with common nails, and lengths generally stop at 3½ inches, going up from 2 inches.

Checkered heads are often used to aid paint adhesion, and a double coating of zinc is applied. Zinc coating is compatible with aluminum siding; bare steel nails are not.

Vinyl siding nails are much like wood siding nails with the exception of having a larger (7/16 of an inch most of the time) head to prevent tearing of the vinyl. Both types come plain or with a deformed shank.

Box nails are similar to wood siding nails but they have a modestly thicker shank (11 gauge versus 12 gauge) they are the type recommended by the FHA for plywood sheathing.

Hardboard siding nails have a smaller head than do wood siding nails. They are meant to nail in materials such as Masonite that require a slightly stiffer nail for good penetration. Once started, nails drive easily in Masonite, but the starting shot must be firm, on target, and powerful enough to penetrate immediately.

Cedar shake siding nails begin to bring in real differences with their blunt points (to hold down splitting) and slim shanks tied to a small head (checkered brad head is usual).

Ring shank and straight shanks are available, and the nails must be hot-dipped galvanized or painted before use.

General purpose is another name for common nails, but they usually have a special feature such as a barbed shank. Barbed shank deformation doesn't give quite the holding power of ring and other deformed shank types, but it is less expensive most of the time. These nails are not usually available in lengths over three inches, have checkered heads, and are double-dipped in zinc.

Roofing nails for shingling asphalt-based and asphalt fiberglass-based shingles are designed specifically for the needs of roofers. The large heads prevent or reduce tearing of the asphalt material, and they also provide an easy target for fast nailing. Generally, sizes start at ¾ of an inch (some companies start longer), and go up to 2½ inches. All must be galvanized or aluminum.

The shorter sizes are used on new roofs over clear decks. The longer sizes are used when there is a particularly thick deck to penetrate or a layer of old shingles to get through. Shingle nails must always go through the sheathing under the shingles.

The *insulation roof deck* nail comes in sizes starting at 3 inches. Penetration should be through the roofing material, through the insulation, and through the sheathing. As always, galvanizing or aluminum is essential, and deformed shanks hold far better.

Cedar shingle nails vary from 1¼ inch to 2 inches. The shingles are thinner than shakes most of the time; long nails aren't essential.

Shingling needs to be done with larger heads to prevent lifting of the shingle and splitting, in this case, in high winds. Such nails come only with a straight shank, nice and smooth, from most markers.

Finishing nails and *casing* nails are similar in appearance and design. The only real difference is in the head shape. Finishing nails have a brad head style designed to be countersunk with a nail set and then filled over with wood putty. Casing nails have a countersink only type of head meant for driving just flush with the surface. Both are used for moldings and door and window casings, and sizes generally run from just over 1 inch up to 4 inches. While some companies even make longer ones for heavy outdoor trim work. They are very seldom needed. The nails might be unfinished mild steel, galvanized, or aluminum, depending on whether the use is to be indoors or out.

As a group, *metal roofing* nails are rather interesting. The head designs provide the major differences from other roofing nails, including those intended for most other roof styles. Some metal roofing nails also serve very well for fiberglass reinforced plastic (FRP) materials.

Rubber washer nails are generally available in lengths to 4½ inches—with 1½ inches being about the shortest you're likely to find most places—while silicone washer nails with fat heads come in very similar sizes (with shanks of the same gauge). Almost all are ring shanked, and galvanizing is imperative for any durability.

Lead washer nails usually come in barbed shank and ring shank styles, galvanized of course, in sizes to 2½ inches, with the lead washer close up under the head. Rubber washers are also placed close under the head. Silicone washers tend to ride down the shank a fraction of an inch and are pushed up to the head when the nail is driven. Extra-wide rubber washer nails are handy for both corrugated asphalt roofing, which you and I are unlikely to use, and for FRP materials. The washers ride down from the head and are 9/16 of an inch wide as compared to the standard rubber washer size of less than ⅜ of an inch.

Umbrella-head nails have concave head caps that work to seal the nail hole and have threaded shanks to increase holding power. Lengths again stop at 2½ inches, and the nails are most useful on flat, metal roofing materials. There are also several other forms of lead-head nails, using a lead that's already compressed and forming the head of the nail instead of being set on as a washer. These styles usually stop at 2½ inches long and serve as fasteners for flat metal roofing, but they can also be used on ribbed styles if the nailing is done through the portion of the rib that touches the rafters (many are nailed through the higher portion of the rib).

Masonry nails are hardened and should only be driven with hammers suited to the purpose. Sizes range from ½ inch to 4 inches.

Holding power in masonry is poor with any kind of nail because the nail generally knocks away so much material it has little to grip. Far better results come with the use of other types of fasteners for masonry, starting with driven fasteners using the Remington power gun. This gun uses .22-caliber blanks to fire hardened steel pins or nails into the material.

Because of the power of the cartridge, crumbling is held to a minimum. The sharp pins hold very well compared to any other form of masonry nail. Masonry nails generally are useful for light work, no matter the size of the nail, and should be ignored when heavy holding power is required on any job. I tend not to use them except to hang tools on unfinished basement walls and such places.

Drywall nails have relatively small heads compared to shingle nails, but large heads compared to some other nails. Use only ring-shanked or other deformed shank styles for best holding power. Make certain the nail is at least 4 times as long as the wallboard it must hold is thick. Most often, wallboard in ⅜ and ½ inch thicknesses will be used. Your nails will need to be at least 1½ inches to 2 inches long; 2 inches is often the maximum length you'll find stocked anyway.

Use galvanized styles for high-humidity areas such as bathrooms, home steam rooms, and so on. Drywall nails are always set into place using a hammer or a drywall hatchet that has a belled driving face. This allows you to dimple the drywall without breaking the paper covering the gypsum. This allows for easy filling over the nailhead when the time comes to lay on the joint compound.

Underlayment nails are available with flat and countersunk heads. On first glance, they resemble drywall nails quite closely. Ring shanks should always be used, and galvanized types are useful in areas where water will stand for a long time. Ring shanks are needed on underlayment nails all the time, but require emphasis when resilient tile or sheet vinyl flooring is to be used because popped nails will show nearly right away.

Flooring nails are available in wire and cut styles. The wire styles include a choice of countersunk or casing heads, with spiral shanks. The wire types are available standard mild steel wire— stiffened wire or hardened—with the casing-head wire types used for hand driving and the countersunk styles suitable for machine driving.

You can rent a floor nailing machine if you're installing new floor. It will nail and set the nail in the same operation. Hand nailing requires that you come back with a nail set, usually laying it on its side in the space above the tongue, to set the nails. Flooring nails, when hand driven, are never driven all the way in one operation. You'll always need to leave about ⅛ to ¼ inch for the nail set to keep from mangling the tongue on the strip flooring. It's a good idea to leave the same amount even when you're face nailing. Using a nail set will prevent marring the face of the flooring with hammer marks.

There are many specialty nails. *Square-cap* nails are useful only when applying built-up roofing, and on some sheathing jobs. The one-inch square head is simply not needed elsewhere.

Pallet nails are meant for building and repairing pallets, and also serve well for repairing wagon beds. They aren't too handy in general use.

Nail materials vary with intended uses, but today you'll find most styles available in aluminum as well as mild steel, and most companies offer hardened steel in most nail styles. Hardened steel nails generally are not supposed to be driven with a standard claw hammer because they will fairly quickly ruin the striking face.

A nailing job is far safer if you wear safety glasses, as silly as it may seem at times. With hardened nails, safety glasses are imperative because they can break in two more easily than mild steel nails. The heads can break and they can chip any hammer. For the same reason, safety glasses are needed when you do any masonry nailing. The masonry can chip, the hammer, or any of the nails might chip.

The first time a nail zips by your head, no matter how far off it really is, you'll feel far better for wearing safety glasses. If you do much nailing, sooner or later you'll find the occasional nail or nail part flying around. Near misses cause this and so do the faulty nails that slip through in manufacturing. Knots and other faults in wood can also make nails fly around as they're being driven. Be safe and keep on seeing. Use safety glasses!

Brads are little more than short nails, usually modeled along the lines of finishing nails. In general, brads are 1 inch or less in length, quite slender, and are used to make picture frames, hold light moldings together, and so on. Useful for many small jobs, they're driven with a 13-ounce hammer in most cases so that control is at a peak and you're less likely to miss and strike the light material being fastened.

SELECTING NAILS

Selecting the proper nail is a relatively simple job once you know a little about nails in general, but you need to know one or two other things as well. First, you need to know the size of the member to be joined. It's also best to know the size of the stationary member, or the member to which you're planning to join another so you can select for nail length. You also need to know whether or not the nail will be used indoors or out, and the type of head and point most suitable to the job.

The style of shank is not irrelevant, but in virtually every case you'll find one of the deformed shank styles provides far greater holding power than does a smooth shank. You'll want to select for the optimum in holding power. Nails hold by forcing wood fibers apart in a wedging action. Those wood fibers will then try to force their way back to their original position, giving the holding power. If portions of the nail shank overlap the fibers after this action, obviously holding power will be greater.

The selection of a nailhead is related to the end use of the nail. Headless or small-headed nails don't offer the pull resistance that large head nails do, but they also don't show up as markedly on the finished job. Small heads are easier to countersink and fill for a neat-looking finish. Casing nails stop flush at the surface and offer very little metal in the final appearance of the work. Of course, paneling nails and some molding nails can be color matched to the building material so there is no appearance of metal and no need to set the nails.

For strong fastening, you'll find a large head best. The larger the head the more holding power the rest of the nail can exert. Actually, with a large head the shank of the nail is able to exert its full holding power before you run into problems. Small-headed nails allow the board to slip over the head before the shank holding power is gone. For materials that tear relatively easily, such as asphalt-based shingles, roll roofing, and builder's felt or paper, large heads are imperative to prevent tearing when wind lifts the materials slightly.

Washer heads and umbrella heads seal openings through solid finish materials such as FRP panels and metal roofing. Heads can also be used as decor items. Too often, people will use a cut nail head as a decoration when cut nails were virtually always hidden. More often, wrought nailheads were visible, and Tremont's rose-head styles imitate this look to a degree while also providing their own look for a project. Tremont's Decor Nails, with their ⅝-inch shanks, are also good for this sort of work where holding power isn't needed.

As with all else when choosing nails, suit the head to the job to be done, and the work will last longer and look better as long as it lasts.

In most cases, the point selection is made for you when the nail is produced for a particular job. Occasionally, a nail company will offer a variety of nail points in common or finishing nails (or in other styles). When that happens, you should consider your job requirements carefully before making the selection. A nail point is a strong influence on both nail holding power after it's driven and in wood splitting as the nail is driven. Some woods split far more easily than do others.

Oak splits easily, for example, and the standard diamond-point nail should be either blunted—tap it with a hammer while holding the head on a solid base; tap, don't slam it—or it should not be used.

A blunt diamond or a chisel point is far better. Blunt point nails have more holding power, but they are a bit harder to start and drive than other types of sharp-pointed nails. The diamond point has become the most popular these days for a combination of decent workability with little splitting.

Gypsum wallboard requires a long diamond or a long needle-point nail so that the gypsum between the covering and backing sheets of the wallboard is not crumbled when the nail is driven. Material such as Masonite is most easily fastened using round, needle-point nails that spread the material apart more gently than does the diamond point. This makes starting and driving the nail an easier job. Carpet tacks usually have long, round points.

Wedge or chisel-point nails are best used with most hardwoods. Poplar, basswood, and other soft hardwoods can be nailed using the same points used for firs and pines with no problem. Such nails give excellent holding power and are easy to drive even when you have to use fairly large nails. When you have a lot of thick wood to drive through, consider the chisel point for all types of wood. It is just about the all-around-easiest-to-drive nail going.

Shank shape plays an important part in nail-holding power. The smooth, round shank of today's common nail provides the least holding power of all generally available nails. The shank shape of cut nails, either triangular or rectangular or square, provides the greatest holding power (about 70 percent greater than a round, smooth shank of similar size, according to Tremont Nail Company) without use of a deformed shank.

Deformed shank nails are the modern way to go for greatest holding power. Essentially, moving to a *barbed shank* nail will provide a small step up from round shank nails. The angled or horizontal barbs hold better than a smooth-shanked nail would but don't begin to approach the holding power of a threaded nail.

Threaded shank nails offer the greatest holding power of any nails. There are, basically, three styles from which you can choose. *Spiral threading* is fairly common in most nail types, and provides you with a nail that turns a bit as it's driven. These work very well in hardwoods and in dense softwoods such as yellow pine because they drive fairly easily, and have optimum holding power.

For aluminum nails, a spiral thread is about the only sensible way to go with deformed shanks. The easier driving that results will give you fewer bent nails.

The annular threaded shank is also called the ring shank and has rings set individually (without the spiral cut to feed one into another). The nail cannot turn as it's driven so the wood fibers are forced over the ring shoulders and into the grooves to lock there. These nails are tops for holding power in softwoods and in most medium woods. Don't use them in oaks and hard maple, but expect them to do very well in soft maples and other such woods. They drive easily into knots (comparatively speaking, that is—nothing really drives easily into knots), and are excellent for use on stairs and floors to help prevent squeaks.

Used with tongue and groove material anywhere, the ring-shanked nail is good at drawing the materials together. Any board warping with tongue and groove—and there's always some in my experience—and you've got a job getting things inserted as tightly as needed for a good finish.

You'll find combinations of annular shank nails with spiral-threaded nails. These offer some of the features of each nail type, presumably with a lock on the advantages and no worries about the disadvantages.

The *knurled thread* nail is often called spiral threading by manufacturers, and uses a vertical thread with open flutes to give extra holding power. It's most often used for masonry nails because it cuts down somewhat on cracking and dusting of mortar and concrete while providing good holding power. You'll sometimes find the style used in large nails meant for heavy-duty use in woods, but the primary aim is for masonry. Most knurled thread nails are hardened.

Nail materials and *finishes* are made of mild steel or aluminum. Mild steel nails will be stiffened (can be nailed with claw hammer) or hardened (should be nailed with a hand drilling or engineer's hammer).

Whenever aluminum is used in a job, you'll find it best to use aluminum nails. If other metals are used, bimetallic corrosive action will be great, and the nails won't last. Aluminum nails are also suitable for most jobs where unhardened mild steel nails can be used, with or without the steel nails being galvanized, because corrosion is not a problem.

Aluminum nails do tend to bend more easily than do steel nails, but in most uses this is of little importance. Cost is somewhat higher but not as bad as it often seems because there are many more aluminum nails per pound than there are steel nails per pound. The higher per-pound cost is often at least partially offset by the greater quantity.

Zinc or *galvanized* finishes are probably the most common coatings to prevent or reduce corrosion problems with steel nails. Galvanizing is generally done in a tumbler producing a thin, gray coat of zinc suitable for moderate outdoor exposure. Electroplated zinc coatings lay on only a thin, shiny coating not really suitable for outdoor use. Mechanical plating is done by rolling the nails in a barrel with small glass pellets and zinc dust, with an activator fluid to help the process. The finish is similar to galvanized and is not much better.

Many companies call their processes hot-galvanizing. This is often confused with hot-dipping. Hot-dipping is the premium way, both in cost and durability, to coat nails with zinc. Hot-galvanizing improves only a little over cold-galvanizing in a tumbler.

Maze produces a nail, the Stormguard, that is double hot-dipped. As always, you pay for the feature, but with some such nails in service for more than 60 years, there is a case, and a solid one, to be made for using such top-quality nails.

Various other methods for coloring or coating mild steel nails include blued finish (done by heat-treating the nail and suitable for temporary outdoor use only) and painting. Painted nails, if the heads are to show, are best driven with a hammer using a plastic cap to keep from marring the nails. Outdoor durability is not nearly as great as with hot-dipped nails. Some consideration should be given to that factor when considering nails. Aluminum nails are available with painted heads, if a color match is required, and corrosion is no problem. The extra cost is minimal.

Driving nails is not a complex task, but some tips can aid you when it comes time to get maximum holding power and minimum nail bending and wood splitting. First, masonry nails are not meant to be driven more than ¾ of an inch into the masonry material. Longer nails are meant to support items held to the masonry with that extra length.

From that point, remember to drive nails at a slight angle towards each other when space permits. This tends to increase holding power. Toenailing is best done from both sides of the T joint made, and is most easily done from the first side if you brace the bottom of the material. A stud, being toenailed to a floor or sole plate is an example. This helps prevent sideways movement.

Nails driven directly into wood with the load placed to pull along the length of the nails can only be loaded lightly. The closer the loading gets to a right angle the stronger your joint. Nails driven with the grain do not hold as well as nails driven across the grain.

Nail length in the finished joint is important. When standard nailing is done, the nail in the stationary member should be two times as long as the nailed-on member is thick. Nails placed too close to end grains will almost always split the wood, unless you drill pilot holes, as will nails placed too close to board edges. If the wood is particularly expensive, remember it only takes a second or two, even with a hand drill, to set a pilot hole.

Nails placed on the same grain line, and too close together, are liable to also cause splits. It's better to set one or the other a bit to the side so the nails are staggered. Pilot holes for nailing are most often needed in hardwoods. You can use a headless brad or a drill bit about ⅓ the diameter of the nail shank. Never make the pilot hole more than two-thirds the size of the shank's diameter.

Nails can be started and driven with one hand. Either insert the nail into the claw so the nailhead rests against the base of the eye or hold the nail against the side of the hammer with your hand. Start the nail and drive as you would normally drive. The first technique only works if you are using a curved claw hammer. In both cases, a good sharp rap of the hammer should start the nail. At times you'll have some problems with especially hard-surface materials. Too many shots to the thumb over a period of time will make anyone flinch. The nail and brad holder called Molly Thumb-saver will take a ½-inch brad or a 10d nail. The Thumb-saver has a spring-loaded slide that is depressed and the nail is inserted into the exposed slot. Give the nail a shot, depress the slide clamp, and the nail is started. Not much chance of hitting your hand there. The tool is great for smaller nail starts where you'd be tapping your fingers fairly often simply because the nail size almost makes it necessary.

A hammer is held in a comfortable grip that should have the edge of your hand even with the bottom of the hammer grip, and arced down from the shoulder to strike the nail squarely on its head. A full shoulder swing isn't always needed. Your aim will be improved if you start the nail, then rest the hammer head lightly on its head before raising it. Lift the hammer from the nailhead, and give a firm blow.

If the nail starts in at an angle you don't intend or starts to bend, pull it and use a fresh nail. It's at this point the claw on the hammer (Fig. 2-19) is useful. The partially driven nail will pull with some ease.

Fig. 2-19. Pulling nails with a hammer. Courtesy of the Stanley Works.

SPECIALTY NAILS

There are many special nails and nailing tools that can provide a lot of help during building and remodeling projects of many kinds. This tool is simple to use, but is moderately costly in use—the tool itself is almost always available somewhere on sale for under $30, but the loads and pins tend to add up, though I must admit the savings in work is considerable. I once objected, in a test article on the Remington Power Hammer, to the cost of the varied blank loads,

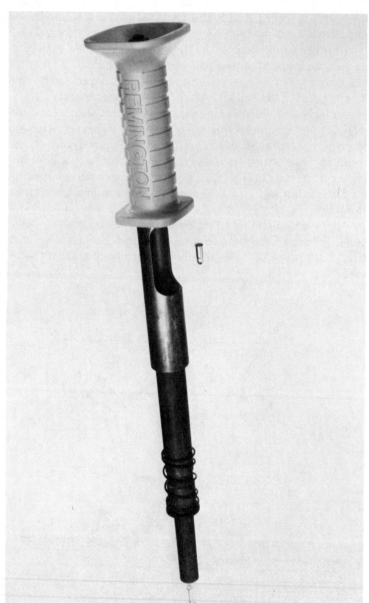

Fig. 2-20. Remington power hammer. Courtesy of DRI, Inc.

and still feel the same way a few years later. When .22 ammunition costs so little, why should these blasted blanks cost three to four times as much? They do, though.

The Remington Power Hammer shown in Fig. 2-20 is from DRI Industries. The Power Hammer is not unusual in operating principals for the construction industry. Larger models have been in use for many years, but the cost of the larger tool and the loads is prohibitive for most do-it-yourselfers.

The Power Hammer won't fire unless you've got the tip placed against some solid surface, pushed down, and a load in place. You start by inserting the pin, or power nail as the company calls it, in the nose, or muzzle, of the Power Hammer. Pull the muzzle forward and you can then insert the power load in the chamber. Slide the muzzle part way back to close the chamber, and then place the muzzle against the work being attached, push down to compress the spring, and give the end of the Power Hammer a sharp rap with a hammer. The load fires, the pin is driven, and you've got a good fastening. See Fig. 2-21.

OPERATING INSTRUCTIONS

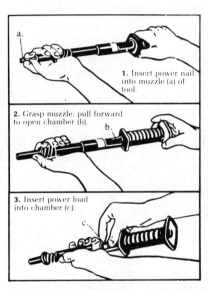

1. Insert power nail into muzzle (a) of tool.

2. Grasp muzzle, pull forward to open chamber (b).

3. Insert power load into chamber (c).

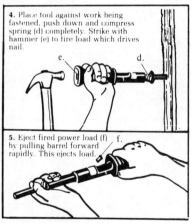

4. Place tool against work being fastened, push down and compress spring (d) completely. Strike with hammer (e) to fire load which drives nail.

5. Eject fired power load (f) by pulling barrel forward rapidly. This ejects load.

Fig. 2-21. *Power nails application.* Courtesy of DRI, Inc.

		1"	Electric box to cement block. Mail box to cement block. Electric box to concrete. Shelf bracket to concrete.
		1 ."	1" x 2" wood furring to cement block. Awnings and shutters to horizontal mortar joints. 1" x 2" wood furring to concrete.
		2 ."	2" x 4" wood to concrete. 2" x 4" wood to cement block.

POWER NAILS APPLICATION CHART

I've used the Power Hammer to install furring, studs, electrical boxes, metal signs, and a number of other items to concrete block, concrete floors, and brick mortar joints.

Loads and nails, or pins, are sold separately, with different length nails for varied jobs, and different amounts of powder in the loads to provide the correct power—some of the extra heavy loads would probably run the short nails an inch or so below the surface of most concretes.

Pins, or nails, are available in lengths of 1 inch, 1½ inches, and 2½ inches from DRI Industries. A catalog can be obtained by dropping a card to DRI Industries, 11100 Hampshire Avenue South, Bloomington, MN 55438.

Loads available are medium, for the short nails, heavy for the 1½ inch nails, and extra heavy loads for the 2½ inch nails. Loads run just under $9 per hundred, pins vary in price according to length (just under $9 for 1-inch nails to about $15 for 2½ inches, per 100). A special hardened steel is used, and I know of no company selling loads or nails that are made by anyone other than Remington.

Consider costs carefully and consider renting a larger, more powerful (and faster-loading) tool if you have a moderate-to-large job to handle. Commercial versions use .25-caliber or .38-caliber loads, and will drive nails through steel, as well as concrete. The loads cost more and the tools are very high priced; rental is the only answer.

Tacks (not thumbtacks) are the next item on the driven fas-

Fig. 2-22. Tack sizes and counts. Courtesy of Hillwood Mfg. Co.

STANDARD TACK SIZES

ESTABLISHED LENGTHS OF TACKS MEASURED UNDER THE HEAD

Nos.	1	1-1/2	2	2-1/2	3	4	6	8	10	12	14	16	18	20	22	24
Ins.	3/16''	7/32''	1/4''	5/16''	3/8''	7/16''	1/2''	9/16''	5/8''	11/16''	3/4''	13/16''	7/8''	15/16''	1''	1-1/8''

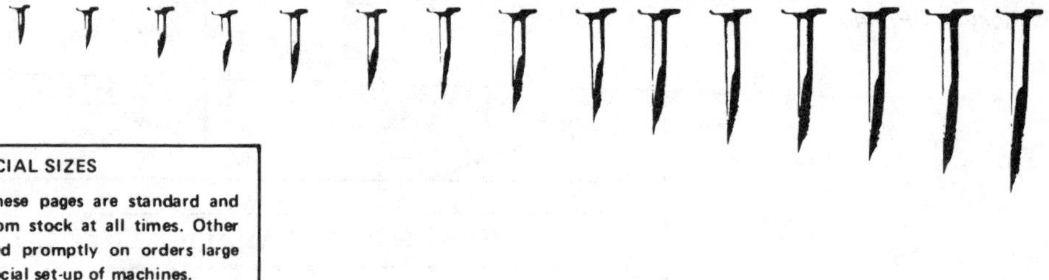

SPECIAL SIZES

Sizes indicated on these pages are standard and normally available from stock at all times. Other sizes can be furnished promptly on orders large enough to warrant special set-up of machines.

STANDARD COUNTS PER POUND

Ounce	1-1/2	2	2-1/2	3	4	6	8	10	12	14	16	18	20	22	24
Carpet	.....	.....	.....	1904	1600	1248	1104	880	.....	.....	.....	.....	.....	.....	.....
Upholsterers	7328	5600	4032	3000	2400	1760	1440	1200	1040	880	720	640	576	512	440
Trimmers	8000	6400	5600	4000	3008	2640	1792	1440	1200	1024	896	.....	.....	.....	.....
Bill Posters	.....	.....	.....	1264	960	608	544	496	.....	.....	.....	.....	.....	.....	.....
Gimp	.....	.....	5000	3488	2992	2496	1840	1600	.....	.....	.....	.....	.....	.....	.....

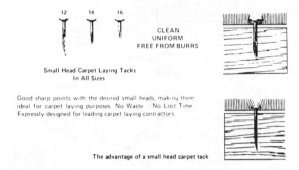

HILLWOOD BLUED, SMALL HEAD
CARPET LAYING TACKS

Satisfy the most exacting users

12 14 16

CLEAN
UNIFORM
FREE FROM BURRS

Small Head Carpet Laying Tacks
In All Sizes

Good sharp points with the desired small heads, making them
ideal for carpet laying purposes. No Waste — No Lost Time.
Expressly designed for leading carpet laying contractors.

The advantage of a small head carpet tack

*Fig. 2-23. Tacks and their uses.
Courtesy of Hillwood Mfg. Co.*

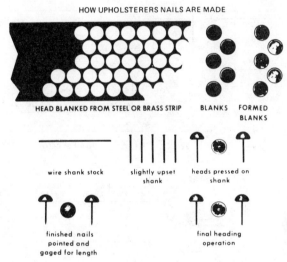

HOW UPHOLSTERERS NAILS ARE MADE

HEAD BLANKED FROM STEEL OR BRASS STRIP   BLANKS   FORMED
BLANKS

wire shank stock   slightly upset
shank   heads pressed on
shank

finished nails
pointed and
gaged for length   final heading
operation

tener agenda. Tacks come in a great many other sizes, styles and uses, and some are exceptionally handy to have around the house. See Fig. 2-22.

Carpet tacks from Hillwood Manufacturing Company offer a needlepoint with a very sharp tip, and a wide range of materials and lengths. See Figs. 2-23 and 2-24. The lengths range from 3/16 of an inch to 1⅛ inches. The tacks are made of steel, copper, brass, or aluminum. With the moderately long #18, there are 640 or so tacks per pound.

Upholsterer's tacks differ little from carpet tacks. Versions include basket, bill poster, copper cut, hide tacks (steeper shank and larger head), webbing (with and without barbs), screening aluminum and copper, usually), and leatherhead carpet. In addition, there are decorative upholsterer's tacks with a wide variety of head

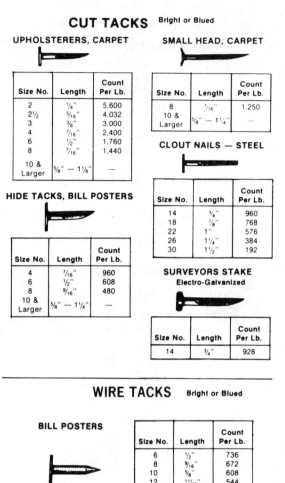

CUT TACKS Bright or Blued

UPHOLSTERERS, CARPET

Size No.	Length	Count Per Lb.
2	1/4"	5,600
2½	5/16"	4,032
3	3/8"	3,000
4	7/16"	2,400
6	1/2"	1,760
8	7/16"	1,440
10 & Larger	5/8" — 1⅛"	—

SMALL HEAD, CARPET

Size No.	Length	Count Per Lb.
8	7/16"	1,250
10 & Larger	5/8" — 1¼"	—

HIDE TACKS, BILL POSTERS

Size No.	Length	Count Per Lb.
4	7/16"	960
6	1/2"	608
8	9/16"	480
10 & Larger	5/8" — 1¼"	—

CLOUT NAILS — STEEL

Size No.	Length	Count Per Lb.
14	3/4"	960
18	7/8"	768
22	1"	576
26	1¼"	384
30	1½"	192

SURVEYORS STAKE
Electro-Galvanized

Size No.	Length	Count Per Lb.
14	3/4"	928

Fig. 2-24. Tacks and their uses. Courtesy of Hillwood Mfg. Co.

WIRE TACKS Bright or Blued

BILL POSTERS

Size No.	Length	Count Per Lb.
6	1/2"	736
8	9/16"	672
10	5/8"	608
12	11/16"	544

styles that are left to show. Generally, decorative nails and decorative tacks come with round, pebbled heads, a flat flower design and a round flower design. Plain round tops are available in various materials of decorative value. Cost is usually moderate (DRI is currently asking about $10 for a kit of some 564 decorative tacks and nails).

Driving is usually done with a tack hammer using a light head weight (most common are 5 and 11 ounce), with magnetized heads making that first starting tap easier when you use steel tacks. Otherwise, consider using Molly Thumb-saver (Fig. 2-25).

Corrugated joint nails (Figs. 2-26 and 2-27) are often used to nail miters in wooden door frames and, from behind, in casings. Corrugated fasteners are a problem when it comes to splitting wood. This is in part because the fastener itself is long, but it's mostly because the fastener is meant to be driven at the corners of

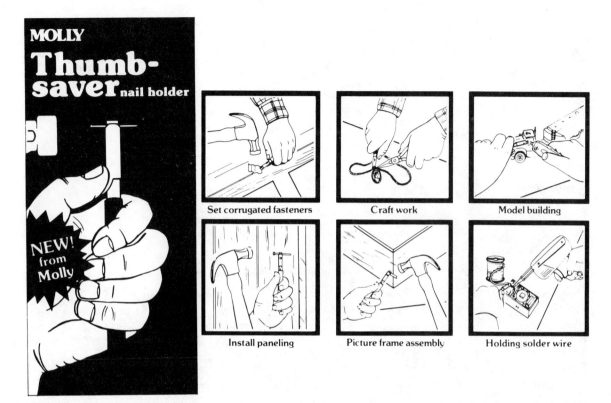

Fig. 2-25. The Molly Thumb-saver.
Courtesy of Bostik.

the wood, with the grain, which is a sure way of having split problems. I keep some ½-inch-by-5 Hillwood joint fasteners on hand for picture-frame repair and molding repair. The 5 stands for the width, which is five corrugations, (in this case, just over 1 inch). Such fasteners are easily trimmed with wire cutters or heavy tin snips.

Molly Drivetool (Fig. 2-28) and Drivepins (Fig. 2-29) are used for fastening furring, 2 × 4s, and other materials to masonry surfaces wherever needed. You've got a choice of pins or studs. The studs offer a ¼-by-20 thread on the undriven end. The thread is ½ inch long and there are lengths of ½ inch, ¾ inch, and 1¼ inches for the driven shaft. These can be very handy for attaching secondary items to furring, etc.

The Drivepins come in lengths from ¾ inches to 3 inches. The Molly Drivetool is a simple item of plastic and metal. The Drivepin is placed in the end (the "muzzle" that fits against the wood being fastened to the masonry). You then tap the tip of the Drivepin to knock the lock guide washer into place and set the Drivetool at a right angle to the work. Use a hand-drilling hammer, preferably, to smack the fastener top of the Drivetool and everything's in place.

The Drivetool provides you with suitable pins that are held solidly in place while you hammer. With less wiggle on the fastener,

Fig. 2-26. Hillwood corrugated nails.

you get fewer chips and dust debris from the masonry and a stronger joint because there is more surface left to grip the pin. There is no need for drilling in many cases, and is far less expensive than the Power Hammer. It's also slower to use. There's sufficient holding power for furring, studs, paneling, and most light-duty to low-medium-duty brackets. The Power Hammer will give you much the same, with more holding power for medium-duty bracketing. In both cases, you must use the special fastening pins provided by the tool manufacturer in order to get sensible results.

Staples (Fig. 2-30) are driven fasteners used to attach materials to, most often, wood. Staple lengths, as recommended by USM, are ¼-inch legs for light upholstery, shelf-paper screens, and shade attachment (window shade); 5/16-inch legs use for thin insulation, plastic storm window installation, general upholstery, and heavy fabrics; ⅜-inch legs are for weatherstripping (well, some types), roofing paper, light insulation, electrical wires, and wire mesh

Fig. 2-27. Corrugated nail set. Courtesy of The Stanley Works.

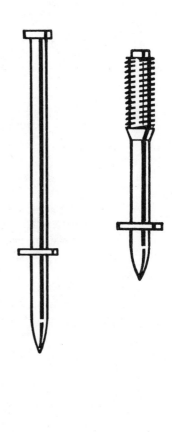

Fig. 2-29. Molly Drivepins. Courtesy of Bostik.

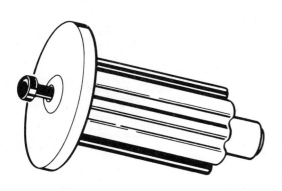

Fig. 2-28. Molly Drivetool. Courtesy of Bostik.

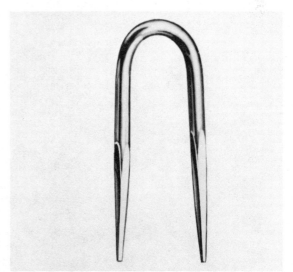

Fig. 2-30. Staple, for hammer driving. Courtesy of The Stanley Works.

(hardware cloth); the ½-inch leg is useful for canvas, felt weatherstripping, some underlayment materials, carpets and fiberglass; the 9/16-inch leg holds ceiling tiles, wire fencing, insulation boards, and some metal lath.

These are general suggestions and I would recommend you test things out or go one size larger before embarking on any major project. Electrical wiring can be held in place with staples, but I personally prefer a hammer-driven staple of the type especially made for the purpose. Exceptions are a low-voltage (thus small) wire or a telephone wire. Even then, I prefer to use a specially tipped stapler.

For more extensive stapling jobs, where electric power is available, I have on hand a Black & Decker 9700 electric stapler (Fig. 2-31). For the few times it does jam, it's very easily cleared, and a safety interlock means you've got to really try to fire a staple through a finger.

The electric brad nailer (Fig. 2-32) from Black & Decker takes 1-inch and 1¼-inch brads, available in strips in colors (most often white and brown). It countersinks as it finishes driving the brad, and

Fig. 2-31. Electric stapler. Courtesy of Black & Decker.

Drives 5 staple sizes: 1/4", 5/16", 3/8", 1/2" and 9/16".

Use for upholstery and furniture repair, installing insulation, tile ceilings, installing plastic sheeting for insulations, repair or building screens, tacking art canvas or fabric to frame. Exclusive solid state circuitry, delivers maximum power to stroke. Built-in fabric screen stretcher at front end. Easy back end loading permits loading from any position.
Specs: 120V AC, 10 amps, 1-7/8 lbs. net wt., 2-1/8 lbs. ship wt.

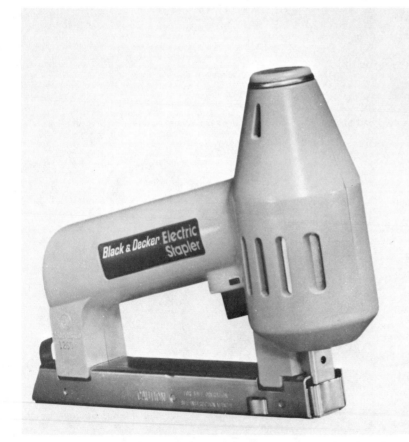

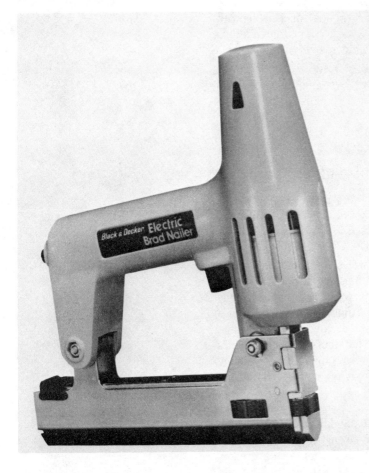

Drives two sizes of brad nails, 1" and 1-¼", in 4 colors: off-white and three shades of brown. Angled base gets in close for nailing in corners and tight spaces. Use for paneling, decorative trim, floor and ceiling moulding, furniture repair, picture frame fastening, construction of shelving, cabinets etc., hobby and crafts (bird houses, models, etc.). Easy back end loading permits loading brads from any position.
Specs: 120V AC, 10 amps, 3-wire 10" cord, 3 lbs. net wt., 3-1/4 lbs. ship wt.

Fig. 2-32. Electric brad gun. Courtesy of Black & Decker.

brads made for it come with blunt points to hold down splitting (after all, a lot of the time the nailer is used on very light materials). The shanks are square to provide good holding power.

The brad nailer is an excellent tool for nailing paneling to studs or furring strips, installing cabinet backs, installing softwood moldings at ceilings and floors, making low-cost storm windows, repairing drawers, or adding molding to drawers to change the look (or to cabinet door fronts, or cabinet fronts). Picture frames are easily made and I'm reasonably sure you can dream up some varied projects on your own.

It is imperative that the brads used in any brad nailer be compatible with that nailer. There's not enough price differential between the one-inch and 1¼-inch brads to worry about, so use the appropriate length for the job.

One of the nicest features of the brad nailer is that it nails where there's no room to swing a hammer. The Black & Decker is about 8½ inches high and 9 inches long. It will fit into some pretty tight spots where you'd only be able to get about a 3-inch swing with even a 13-ounce hammer. If your work involves installing ceiling molding in several rooms, the use of a hammer is very tiring. This is especially true when you have to come back and set all the nails, taking care not to swat the woodwork. The same holds true for using an electric stapler for ceiling tile.

WOOD SCREWS

Screws for wood go almost as far back as nails, but only in the past century or so have they become sensible alternatives for a large number of jobs.

Screws are often used to fasten wood to wood or to other materials because they offer several advantages over nails. In some cases, those advantages are so great that jobs traditionally done with nails are now done with screws.

I have here a good-sized box of drywall screws, to be set with a variable-speed electric drill and a #2 Phillips bit, that do a better job in less time. Shopmate screws are great for repairs, and for covering small rooms and such, though cost is higher than for wallboard screw-shank nails. As far as I can determine, the cost is reasonable for the time and labor saved. I've found these drywall screws invaluable when I'd otherwise have to drive many nails over my head. It is a lot faster and easier to simply bit a screw and let the screw be driven by a small electric drill.

In general, the disadvantages of screws come from longer installation time and greater cost of the individual fasteners (though fewer will be needed for even greater strength), plus the added tools required for many jobs. The advantages begin with greater holding power, especially when the load stress runs parallel to the shank of the fasteners. Nails pull out quickly in such situations; screws will hold if properly installed.

Work that has been fastened with screws won't loosen with vibration and on-and-off stress cycles. If there's any chance you'll ever want to take a section apart, screws make that possible without destroying the primary materials. When nails are pulled, the surfaces will be marred badly. Screws also serve to draw items together as they're driven. There's less chance of damage when you use a screwdriver than when you use a hammer (less likelihood of marring a finished piece). Screws also can give a neater appearance, and, like some types of nails, fit well into many decorative schemes.

Wood screws come in a variety of materials. Mild steel, brass, and zinc-plated and cadmium-plated screws are available for general uses. Chrome-plated and nickel-plated screws are used where importance is given to appearance and rustproofing at the same time. The cost of brass and its chances of matching the project are not reasonable.

Fig. 2-33. Right: Common screw head styles and sizes. Courtesy of The Stanley Works.

SELECT THE RIGHT SCREWDRIVER FOR THE JOB.

Place Head or Shank of Screw Over Illustration to Determine Actual Screw Size.

 SLOTTED   PHILLIPS   POZIDRIV†   CLUTCH HEAD   ROBERTSON TYPE   TORX*

FOR SLOTTED SCREWS USE STANLEY SCREWDRIVER		SCREW SIZE	HEAD SIZE	SHANK SIZE	PHILLIPS POINT SIZE	FOR PHILLIPS SCREWS USE STANLEY SCR. DR.	
BIT. NO.	SCREWDRIVER CAT. NO.		ACTUAL SIZE		ACTUAL	SCREWDRIVER CAT. NO.	BIT NO.
	66-117	0			POINT NO. 0	64-100, 64-170	
	63-310, 66-101, 66-102, 66-103	1					
		2			POINT NO. 1	64-101, 64-171, 65-201, 65-206, 65-321, 65-601	64-021 68-301 68-311 68-351
68-353 68-354	63-316, 66-110-10, 66-112-2, 66-113-3, 66-114-4, 66-116-6, 66-118-8 66-140	3					
		4					
68-355	67-130-10, 67-132-2, 67-134-4, 67-136-6, 67-138-8	5			POINT NO. 2	64-102 64-105 64-172 64-175 64-272 65-151 65-202 65-205 65-322 65-342 65-602 65-605	64-022 68-302 68-312 68-352
		6					
66-023	63-383-3, 63-386-6, 66-180-10, 66-182-12, 66-183-3, 66-186-6, 66-188-8, 66-453-3, 66-456-6, 66-458-8, 67-263-3, 67-283-3, 67-286-6, 67-288, 67-683-3, 67-686-6 67-688-8	W7					
		8					
		W9					
68-304 68-356	63-363-3, 66-003, 66-163-3, 66-165 67-663-3	10			POINT NO. 3		
68-305 68-314	67-144-4, 67-146	12					
66-024 68-306 68-315 68-357	63-312, 63-364-4, 63-374-4, 63-396-6, 66-004-4, 66-161, 66-164-4, 66-174-4, 66-204-4, 66-674-4, 67-261, 67-274-4, 67-661, 67-664-4	W14				64-103 64-273 65-203 65-323 65-603	64-023 68-303 68-313
66-025 68-316	63-366-6, 63-376-6, 66-006-6, 66-166-6, 66-176-6, 67-266-6, 67-276-6, 67-666-6, 67-676-6	M 1/4					
		W16					
66-026	63-368-8, 63-376-8, 66-008-8, 66-168-8, 66-178-8, 67-278-8, 67-668-8, 67-678-8	W18			POINT NO. 4		
66-028	66-160-10, 66-162-12						
	66-010-10, 66-170-10, 66-172-12	M5/16				64-104	
		W20					
		W24 & M 3/8					

†Registered Trademark of Phillips Screw Co.

*Registered Trademark of Textron.

*Indicates both wood screw and machine screw size, except where indicated by "M" machine screw only—"W" wood screw only.

WORK SAFELY WITH TOOLS BY WEARING SAFETY GOGGLES

Copper-plated screws are often used to give an antique appearance, as are screws finished in black and various so-called antique finishes, while aluminum screws are readily available for use in projects where the main metal is aluminum and the need for rustproof screws is present. Aluminum screws cost a little more than mild steel. Silicone bronze screws are costly and generally applicable only to marine uses.

Steel screws are usually the strongest. Brass and aluminum hold up far better in outdoor uses, and in certain applications seem inevitable. As an example, oaks contain an acid so the use of mild steel screws is not a good idea. Even when used indoors, the acid can cause staining.

Wood screwhead styles vary more than do screw threads and shanks, but are not nearly as varied as metal and general-purpose screwheads (Fig. 2-33). All wood screws are made up of three basic parts: the head is formed to fit the driving tool tip, and directly under the head is a smooth shank that tapers to a point through the threaded portion to the tip. Most often, wood screws are simply classified as to material, head style, and size. The most popular and readily available wood screws coming in flat-slotted, round-slotted, or oval-slotted heads, with a similar configuration in the Phillips head.

Other recessed-slot wood screws are available, but the Phillips is currently the most popular. Many companies now seem to be promoting a screwhead type that is recessed and square, very similar to the Robertson type used.

Slotted heads are good in low-torque applications where you don't need to apply a lot of power to the screwdriver to draw materials together. Phillips heads allow far greater torque applications, with their crossed, recessed heads. When wood really needs drawing up, they're the best way to go. See Figs. 2-34 through 2-36.

Recessed heads in general allow you to apply much greater torque because there is much more surface area for the driving tool to work against, and because the recession of the head slot leaves sides to help prevent screwdriver slip. Screwdriver slip is the bugaboo that visits all of us eventually. It can mean anything from a badly marred surface when you need to finish it smoothly with a clear finish to badly cut up or banged up hands.

One of the more useful tools when many screws must be driven without power is the Yankee spiral ratchet screwdriver from Stan-

Fig. 2-34. Phillips, Reed and Prince and Torque set heads.

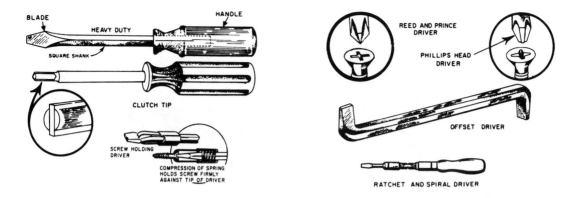

Fig. 2-35. Screwdriver types.

Fig. 2-36. Reed and Prince and Phillips screws.

ley Tools (Fig. 2-37). I've used it to install all kinds of items from toolbox (carpenter's style) bottoms and sides to sliding patio door frames. They are time-savers and work-savers where there's space for use. The standard duty Yankee screwdriver offers a larger selection of bit types, including clutch and Robertson, while basically the heavy-duty model will accept only a Phillips or slotted bit (Fig. 2-38).

Fig. 2-37. The Yankee ratcheting screwdriver. Courtesy of The Stanley Works.

In almost all cases, you'll get the best job with drilled pilot holes for the screws. Such holes are essential in hardwoods. I've installed patio doors without pilot holes and the job isn't difficult because the door frame is drilled and helps to support the starting screw. Otherwise, even toolbox building requires pilot holes. In such cases, I usually arrange to countersink and counterbore using Stanley's Screwmates which do all three jobs in one shot. Chuck them in a drill and mark the depth for the appropriate-size Screw-

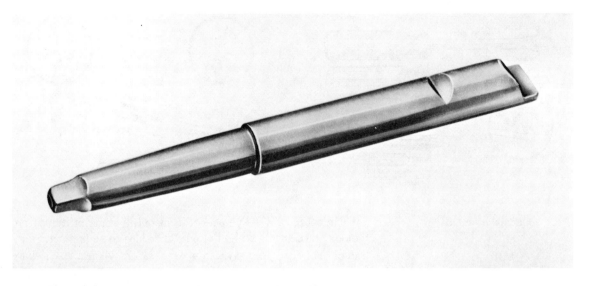

Fig. 2-38. Robertson tip for the Yankee. Courtesy of The Stanley Works.

mate. The set includes #6, #8 (twice), #10, and #12 tools, with
lengths varying from ¾ of an inch to 2 inches (#6 and #12) in ¼-inch
steps. As designated, the screws fit beautifully in the drilled holes,
and you can either leave the countersunk heads visible or fill or plug
those heads that have been counterbored. They save a lot of chuck-
ing time with the drill (if nothing else). My countersink for my bit
and brace works fine, but requires me to remove the pilot hole bit to
use the countersink. This adds to the time needed for installing
screws. While this works without electricity, the Screwmate requires power, but cuts way back on the needs for changing things
around, cutting the time to pilot drill, countersink and counterbore
by a factor of probably 75 percent. A good cordless electric drill
(Fig. 2-39) will take care of power needs if you're away from power
lines.

When *drilling pilot holes* in softwood, drill the hole only half as
deep as the threaded portion of the screw. In hardwoods, you'll want
to drill the hole the same depth as the actual length of the screw. If
you're working with very large screws (#16 and up) in very hard
wood, drill a pilot hole somewhat smaller than the threaded shank of
the screw at its smallest diameter (at or near the tip).

Use a second drill, the same size as the upper shank diameter
(unthreaded portion) of the screws, and enlarge the top of the hole
for about one-third the depth of the hole. This kind of work is far
simpler with Stanley's Screwmates and other special counterbore /
countersink bits.

If screws are still hard to drive, draw the tip and most of the
threaded shank across a dry piece of soap (if the screw is corrosion
resistant) or beeswax (soap's less expensive, but beeswax is nicer).

Use the appropriate screwdriver (Fig. 2-40) for the screw

being driven. Assuming a slotted screwdriver and screwhead, you could easily find yourself working with screws from size #0 through size #24 in widely varied lengths.

Selection of tip size for the screws being driven is crucial. Both too small and too large mean problems. A too small screwdriver tip allows the blade to slip from the screw slot too easily. A too large tip has to be jammed into the screw slot if it will fit at all, and the tip will extend beyond the screwhead sides, marring the work.

Don't start screws with a hammer. Use a screw starter such as Irwin's, a pilot hole, drive a small brad in and pull it out—anything, but don't use a hammer on the screw to get it started. Wood fibers are pushed ahead of the screw when it's hammered, and you may also damage the screwhead so that screwdrivers cannot slot in properly to finish the job.

If you're using the screws on a job with the idea they can be

Fig. 2-39. The Screw Mate in use.

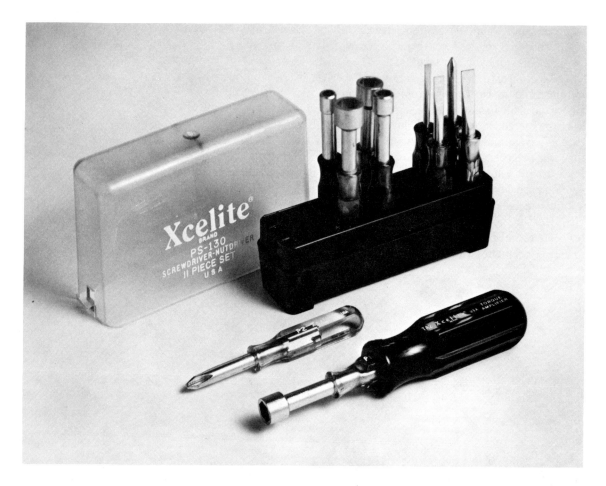

removed and replaced later, remember that extensive use of the item forces the screws to grip hard, and later removal will leave a hole of great size, at least compared to a nail hole. You'll almost certainly find it necessary to fill the hole before returning the same size screw so that grip will be retained.

There are several ways to fill screw holes. You can force some steel wool into the hole or fill the hole with a top grade of wood putty (I like both Elmer's Carpenter's Wood filler or UGL's ZAR Wood Patch for this job, among others).

Don't use plastic plugs driven into the area if you expect a good hold over time. Consider taking a matching, or close, wood dowel, taper it to fit, coat it with Elmer's Cascamite resin glue, or Elmer's resorcinol glue. For really bad-weather use and extreme exposure, nothing beats resorcinol, but it leaves an ugly glue line. Cascamite resin is a one-part-powder that is resistant to water, mold, and so forth. Resorcinol is a two-part-resin adhesive just about totally impervious to anything except the heat of a blowtorch (and it will

Fig. 2-40. Convertible screwdriver set is a good bet if you are short of tool storage space. Courtesy of Xcelite/Cooper Group.

take that for a short time!). Coat the dowel, drive it in place, and then re-run your screw. Longer screws can be used as replacements when there's room in the work. In many cases, the screws you remove have been designed to do a job and provide the ultimate in size the wood in question will accept without allowing the screw to extend beyond the wood, or splitting the wood.

If a lot of removal and return is expected, consider using machine screws and Teenuts. Teenuts are installed in the wood in holes drilled for the purpose. They accept various size screws (but not wood screws). The Teenuts vary from very small to quite large, and are pressed into holes in the wood. I usually use a hammer to tap them in because the prongs, four per nut, are quite long to serve as a combination threaded bushing and washer, allowing disassembly with no worries about overly large holes.

There are other fasteners available for knock-down furniture or furniture that must be disassembled with some frequency, but you'll usually only find such items in the catalogs of speciality mail-order companies that also sell exotic woods and other such items. Various operations are required to successfully use these specialty k-d fasteners, but the resulting speed of disassembly and reassembly is often worth checking them out and then using them.

There are various ways to keep a flathead screw from working loose when it has been installed and the item built has to withstand a lot of stress and movement. You might try using a center punch near the rim of the head, striking it sharply. This should produce a small metal peel that will act as an anchor, preventing any turning.

Lag screws are coarse-threaded wood screws with a bolt head. They're almost never useful in furniture applications and cabinetry, or, for that matter in light work.

Select lag screws with care. Figure for every inch of wood thickness being held in place, you need a screw extending 1½ more inches into the base substance. This isn't always the case, and you'll find a good grip to be had with lag shields in masonry with only 2 or 3 inches of penetration on a ½-inch diameter lag screw.

Lag screws come in diameters from ¼ inch to 1 inch, with the usual stops between being ⅜ inch, 7/16 inch, ½ inch, ⅝ inch, ¾ of an inch and ⅞ of an inch. Lengths range from 1 inch to 16 inches or more.

To install a lag screw, select a drill bit slightly smaller than the shank diameter and drill a hole to just the depth the screw will penetrate (or a touch less). Drill a second, larger hole equal to the length of the unthreaded portion of the shank. If the lag screw goes through wood into masonry, as is often the case, measure the lag shield to be used and use a masonry drill to make a hole into which the lead shield, or anchor, can be driven (use a soft hammer, or very, very light taps with a claw hammer to set the shield in place.

Try not to deform the top lips but don't get upset if the shield closes a bit as it's driven in (it should).

The hole drilled in the wood should be the same size or slightly larger than the unthreaded shank portion of the lag screw. Coat with wax on the threads and use a socket (adjustable or other) wrench to run in the lag screw. If you expect much stress, either from vibration or other causes, at the screw, insert a washer under the head before screwing it down. This keeps the head, which is relatively small in relation to the shank, from tearing up the wood underneath.

Lag screws may have either hexagon heads (the easiest to drive with socket wrenches) or square heads (the easiest to drive with open end or adjustable wrenches). Use hexagon heads if you must drive a lot of lag screws. A socket and ratchet saves a lot of effort when a dozen, two dozen, or more must be driven.

Wood bolts come in two basic styles. The *carriage bolt* is one and the *machine bolt* the other. The carriage bolt is designed to fit a hole and grip the hole with a portion of the bolt neck so that you can then turn the nut on from just one side, without having to hold the bolt head in a wrench. Neck styles most commonly available are the square and the finned (which is squared with fins from the squared corners of the neck to the bottom of the bolt head). Ribbed carriage bolts also are very handy, if a bit harder to find much of the time (Fig. 2-41).

Wood bolts of the above type are used in holes made just a touch larger than bolt shank size, including the threaded portions. I always use a washer under the nut, but, of course, one cannot be used under the head. Wood bolts of this type make for great strength and fast assembly and disassembly. Machine bolts can be used with washers under both head and nut. Stove bolts are not as precisely made as machine bolts. The price is somewhat lower if a larger number must be used.

Carriage bolts are threaded part of the way up the shaft. Holes are bored for a tight fit, either the same size as the shank or just a hair larger, and the bolts are driven through the hole with a soft hammer (you can use a ball peen or a claw hammer as well). If the surface is to have a good finish, I use a lignum vitae mallet, a rawhide mallet, or a plastic-headed hammer even when final finish doesn't seem all that important (in deck construction, for instance).

Fig. 2-41. Carriage bolt. Courtesy of DRI, Inc.

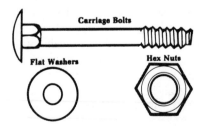

Carriage Bolts

Flat Washers   Hex Nuts

Machine bolts are threaded part way up the shank in about the same manner as carriage bolts, but they do not have the shouldered head to grip wood surfaces the carriage bolt has. The head and nut must be accessible when you're tightening them. You'll find machine bolts with either National Fine or National Coarse threads. You must get a matching nut for the threads, but it will help in some jobs if you remember that fine-cut threads have greater holding power than do coarse-cut threads for the same shank diameter and length. If the machine bolts are 6 inches or shorter, the threads will be cut twice the shank diameter, plus a quarter inch up the shank length, and will add another quarter inch for bolts over 6 inches in length.

These bolts are precisely made, and they can also be used in metal-to-metal applications, as well as wood-to-wood and wood-to-metal. Heads are available square, hexagon, rounded, and in a flat countersink style (hard to find, and almost as hard to use). The nut will generally have the same shape as the head, and the same size, to make matching wrenches to the job a bit easier. They're generally available in hardware stores in lengths to a foot or more, and diameters to ¾ of an inch.

When using machine bolts, bore the hole to the same diameter as the bolt shank, or just a hair larger, and drive the bolt through as you would the corresponding carriage bolt.

Stove bolts are less precisely made than machine bolts so they cost less. In addition, they are really screws with square nuts. The heads will be countersunk style, slotted, or round slotted. Applications are normally metal-to-metal, but use in wood-to-metal and wood-to-wood areas is also possible. With the roundheads, use washers under the head and the nut. With countersunk styles, use washers under the nut only.

BOLT FASTENING

☐ Use washers whenever possible. Make sure they are of the appropriate size. Lock washers should be used whenever vibration is one of the major stress factors on a bolted joint.

☐ Make sure any bolt used is long enough for the job. The bolt must pass all the way through the work and still leave enough to thread down the nut with at least a partial thread rising from the nut's end. When you are measuring for bolts that will not accept washers at one end, measure from the base of the bolt head to get the proper length. Always add washer thicknesses to the bolt length.

☐ In wood, bolts are best set in holes nearly exactly the same size as the bolt. A soft-faced mallet or hammer will drive them on through readily without marring the work surface.

☐ Size wrenches used carefully for a snug fit on nuts and bolt heads. A loose wrench slips. Adjustable wrenches are always pulled to the side that has the adjustable jaw that helps to keep the jaw from slipping open a little and coming off the nut. The procedure also places the greatest turning stress on the much larger stationary jaw of the wrench.

☐ Keep in mind that when lock washers can't be used, and vibration is expected, a drop or two of a sealant on the threads of the bolt, at the point where the nut will be, will serve almost as well, as will a stroking touch on the thread that comes out of the nut with a fine file. Deforming the thread locks the nut in place.

Hooks and eyes (Figs. 2-42 and 2-43) are commonly manufactured of single pieces of wire stock that are then threaded. There are variations where the eye is attached later. Cup hooks usually have a plate to prevent excessive run-in of the threaded shank, and are driven by hand or with a pair of pliers (depending on both the size of the cup hook and the hardness of the wood to be penetrated). Square bend hooks (Figs. 2-44 and 2-45) are installed in the same manner, but are most often used to hold things horizontally, such as some types of curtain rods, or, in our house, various plant hangers. Metal finishes for such hooks should be appropriate to the planned use. Mild steel in plain finish serves well indoors, but decorative and rust-protecting finishes are available for more specific uses indoors and out.

Hooks and eyes are commonly used to hold wooden doors closed when a ring is also included and some have locking sleeves to prevent inadvertent opening. These are almost always galvanized or zinc-dipped because they are handled a great deal, even when indoors, so that rust from acids and moisture on our hands become a problem.

Wood screw hooks for laundry lines have very sharp points, and generally need a good, solid back-up at least 2 inches thick, and preferably thicker, if long laundry lines are to be used. Use pliers to

Fig. 2-42. Hook and eye. Courtesy of The Stanley Works.

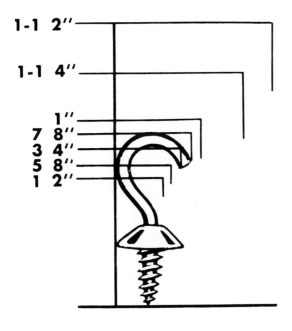

1-1 2″
1-1 4″
1″
7 8″
3 4″
5 8″
1 2″

Fig. 2-43. Cup hooks, sizes. Courtesy of The Stanley Works.

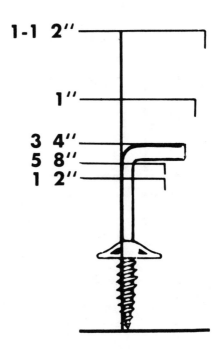

1-1 2″
1″
3 4″
5 8″
1 2″

Fig. 2-45. Square bend hook sizes. Courtesy of The Stanley Works.

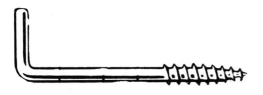

Fig. 2-44. Square bend hook. Courtesy of The Stanley Works.

Fig. 2-46. Laundry line hook. Courtesy of The Stanley Works.

insert such hooks. You'll still find the job a great deal easier if you drill a pilot hole at least half the diameter of the screw-shank portion (Fig. 2-46). Don't thread in beyond the screw shank portion, and do make sure the hook is heavily coated with zinc, or plastic, to protect it from rust.

Chapter 3

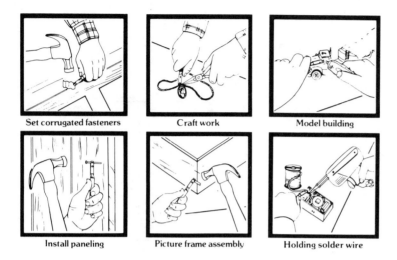

Set corrugated fasteners   Craft work   Model building

Install paneling   Picture frame assembly   Holding solder wire

Wood Joints

Wood joints come in various designs (Figs. 3-1 through 3-62). Each has particular characteristics more suited to some uses than to others. Some can be used as they come from the woodworking shop, with no other fasteners required, while others need the adhesion provided with glues, nails, screws, dowels, staples, or any other wood fastener. Some are very simple to make and use. Others can drive the inexperienced worker up the walls at the design production stages. Without the proper tools, some joints are nearly impossible to make. Some can be made with very simple tools, and some require a great deal of precision fitting and several jigs to get things in the proper order. Some are readily made with an array of minor hand tools. Others require power equipment. If you think you can cut accurate tongue and groove joints with hand tools, you've got my good wishes and sympathy. The job becomes formidable indeed after the first 30 inches or so (about the maximum comfortable single sweep reach with any grooving plane, assuming you're able to locate such a plane today). Mortise and tenon joints are easily made with hand tools, but they can require an awful lot of time if a precise fit is required (as it usually is when such joints are to be employed).

Splitting shingles and shakes by hand is one thing, but making precise woodworking joints by hand is another. Much more skill is required for the joints than to produce shakes, and the development

Structural Joints

Through Dado	End Rabbet	End Lap
Groove	Half Cross Lap	Open Mortise & Tenon
Dovetail Groove	Tongue & Groove	Spline
Drop Leaf	Rabbet	Dovetail Dado

Dado & Rabbet	Dado Tongue & Groove	Blind Mortise & Tenon
Tee Lap	Edge Cross Lap	Half Blind Lap
Half Lap		Through Mortise & Tenon

Fig. 3-1. Structural wood joints. Courtesy of Shopsmith, Inc.

of those skills requires a great deal of time. Not many of us today can run a clean line on the bottom of a mortise using nothing but a mallet and hand chisel. Blind mortises are probably best avoided if joint strength is desired to be optimum.

Basic all-wood joints vary a lot, but most are fairly simple, and relatively easy to make with nothing more than a table saw or radial saw. The more complex joints will require more tools, more accessories, more planning, and more time to produce, but they can be exceptionally valuable in certain uses. The dovetail joint is a good example. Dovetails are altered mortise and tenon joints; the mortise is shaped to take the dovetail shape of the tenon. They can be

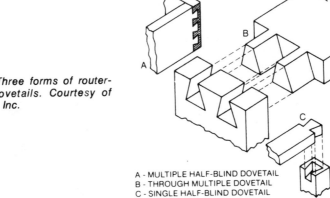

Fig. 3-2. Three forms of router-shaped dovetails. Courtesy of Shopsmith, Inc.

A - MULTIPLE HALF-BLIND DOVETAIL
B - THROUGH MULTIPLE DOVETAIL
C - SINGLE HALF-BLIND DOVETAIL

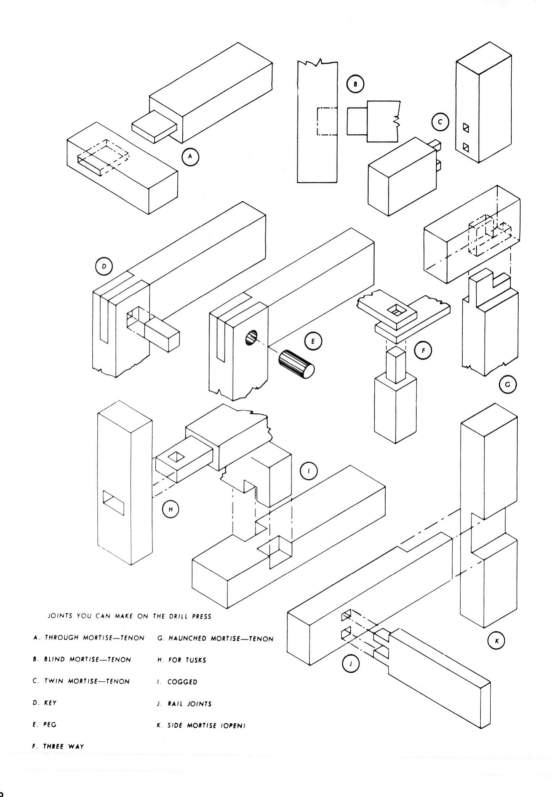

JOINTS YOU CAN MAKE ON THE DRILL PRESS

A. THROUGH MORTISE—TENON   G. HAUNCHED MORTISE—TENON

B. BLIND MORTISE—TENON   H. FOR TUSKS

C. TWIN MORTISE—TENON   I. COGGED

D. KEY   J. RAIL JOINTS

E. PEG   K. SIDE MORTISE (OPEN)

F. THREE WAY

68

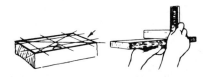

1. WORK FACE

PLANE ONE BROAD SURFACE SMOOTH AND STRAIGHT. TEST IT CROSSWISE, LENGTHWISE, AND FROM CORNER TO CORNER MARK THE WORK FACE X.

4. SECOND END

MEASURE LENGTH AND SCRIBE AROUND THE STOCK A LINE SQUARE TO THE WORK EDGE AND WORK FACE. SAW OFF EXCESS STOCK NEAR THE LINE AND PLANE SMOOTH TO THE SCRIBED LINE. TEST THE SECOND END FROM BOTH THE WORK FACE AND THE WORK EDGE.

2. WORK EDGE

PLANE ONE EDGE SMOOTH, STRAIGHT AND SQUARE TO THE WORK FACE. TEST IT FROM THE WORK FACE. MARK THE WORK EDGE X.

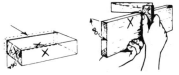

5. SECOND EDGE

FROM THE WORK EDGE GAUGE A LINE FOR WIDTH ON BOTH FACES. PLANE SMOOTH, STRAIGHT, SQUARE AND TO THE GAUGE LINE. TEST THE SECOND EDGE FROM THE WORK FACE.

3. WORK END

PLANE ONE END SMOOTH AND SQUARE. TEST IT FROM THE WORK FACE AND WORK EDGE. MARK THE WORK END X.

6. SECOND FACE

FROM THE WORK FACE GAUGE A LINE FOR THICKNESS AROUND THE STOCK. PLANE THE STOCK TO THE GAUGE LINE. TEST THE SECOND FACE AS THE WORK FACE IS TESTED.

Fig. 3-4. Using a square.

Fig. 3-3. Left: Types of mortise and tenon joints. Courtesy of Shopsmith, Inc.

cut in several ways, but today most are cut as through dovetails because the job is easier that way. Blind or concealed, dovetails require a fair amount of extra touch-up for a really neat, clean fit. The joint is primarily used in furniture building and cabinetmaking.

The tenon is the easiest-to-make part of the mortise and tenon joint. Tenons can be made with simple shoulder cuts, assuming you don't get involved immediately with haunched tenons and such fancy things. Shoulder cuts are made by setting the blade to the exact cut depth and running all four sides of the tenon through the saw. Then the blade is set to cut the length of the tenon. The member is stood on end and all four sides of that end are run through the cut so that the excess material drops off (the cuts can be made in reverse order).

A router will also do the work, but each cut on a shoulder will usually require several passes if the tenon is at all deep. There's no problem at all using a dovetail or tenon saw to make the cuts by hand.

The tenon is easily laid out using nothing more than a try square or combination square and a pencil. Make sure all marks are

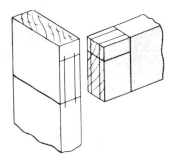

Fig. 3-5. Stub mortise and tenon layout.

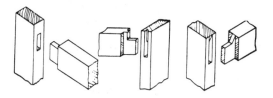

Fig. 3-6. Various mortise and tenon joints: the mortise and tenon is one of the most common wood joints, and has been for centuries.

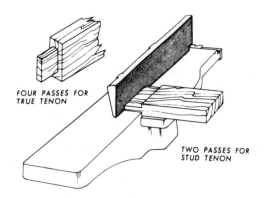

FOUR PASSES FOR TRUE TENON

TWO PASSES FOR STUD TENON

Fig. 3-7. Jointer formed tenons. Courtesy of Shopsmith, Inc.

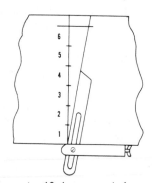

6
5
4
3
2
1

Fig. 3-8. Laying out a 10-degree angle for a dovetail joint.

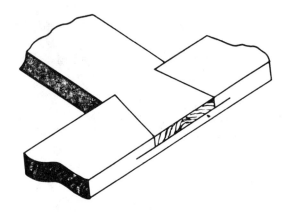

Fig. 3-9. Dovetail half-lap joint.

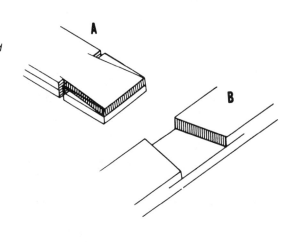

A

B

Fig. 3-10. Marking the dovetail half-lap joint.

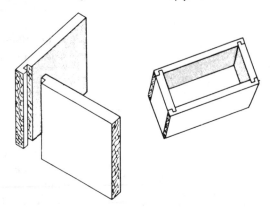

Fig. 3-11. Dado and rabbet joint.

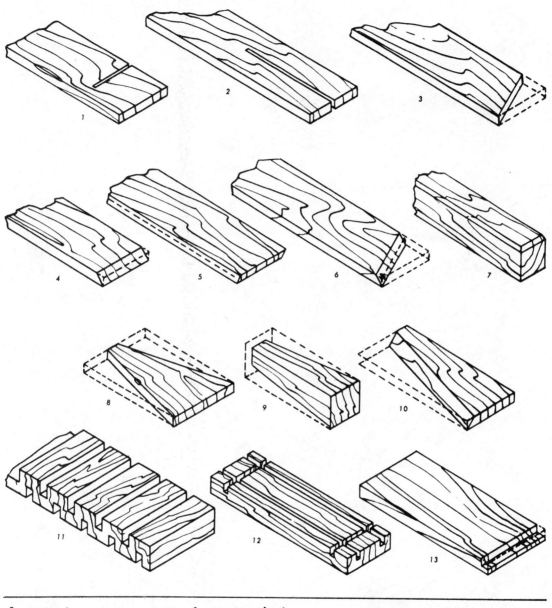

1. crosscut	6. compound miter	10. compound rip-bevel
2. rip cut	7. chamfer	11. kerfing
3. miter		12. kerfing
4. cross-bevel	8. two-sided taper	13. rabbet (two-pass)
5. rip-bevel	9. four-sided taper	

Fig. 3-12. Cuts that are readily made with a plain saw blade. Courtesy of Shopsmith, Inc.

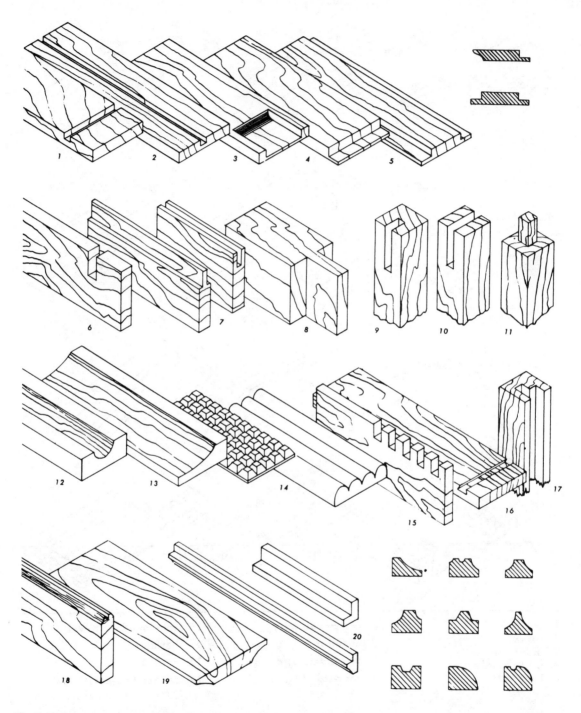

Fig. 3-13. Cuts 1 to 13 are made with a dado blade as are 15 and 17. Cuts 14, 18, 19, and 20 are made with a molding head. The drill press makes number 16. Courtesy of Shopsmith, Inc.

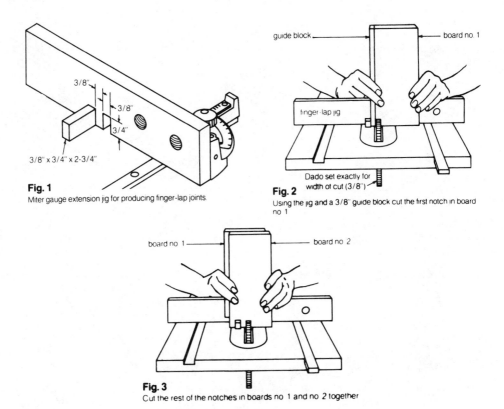

Fig. 1
Miter gauge extension jig for producing finger-lap joints.

3/8"
3/8"
3/4"
3/8" x 3/4" x 2-3/4"

guide block — board no. 1
finger-lap jig
Dado set exactly for width of cut (3/8")
Fig. 2
Using the jig and a 3/8" guide block cut the first notch in board no. 1

board no. 1 — board no. 2
Fig. 3
Cut the rest of the notches in boards no. 1 and no. 2 together

Fig. 3-14. Finger lap notching. Courtesy of Shopsmith, Inc.

accurate. If the tenon is to be blind—that is, to end without coming through the member to which it is attached—it is cut about ¼ of an inch shorter than the mortise depth so that excess glue will have a place to gather. If the tenon is to run all the way through the joint, then it is cut to full length to fit either flush with the opposite side of the mortise or to pass beyond that and be pegged, wedged, or fixed in some other manner.

With a through tenon that just reaches the other side, glue is the most common fastener. Dowels are sometimes run in from the side, through the mortise wall, and through the tenon. Dowels can

2" 1"
SPACES
DRILL CENTER — ¼" HOLES
¼" BOLT — DOWEL
TUBES

Fig. 3-15. Doweling peg. Courtesy of Shopsmith, Inc.

73

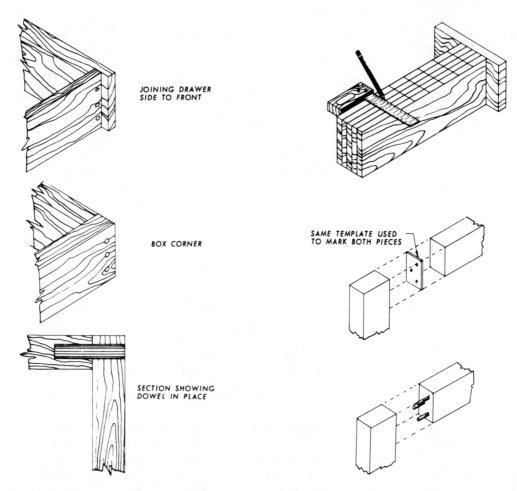

JOINING DRAWER
SIDE TO FRONT

BOX CORNER

SECTION SHOWING
DOWEL IN PLACE

SAME TEMPLATE USED
TO MARK BOTH PIECES

Fig. 3-16. Pegged joints. Courtesy of Shopsmith, Inc.   Fig. 3-17. Pegged joint layout. Courtesy of Shopsmith, Inc.

be combined with glue to give great rigidity and can be used as well with blind tenons.

The mortise adds a bit to measurement needs, and to tool requirements. After marking, the usual first step in producing a mortise is to drill out the excess material inside the marked lines and to the correct depth. The mortise, like the tenon, is either through or blind. Once the excess inner material is removed with a drill bit, either by hand or with power, a chisel is required to square up the corners and remove material left by the drill bit.

If the mortise is to be blind, you'll find it handy to get the bottom as smooth as possible, even though the tenon does not rest on the bottom of the mortise if glue is used. Glue has body so it must be given someplace to go when the joint is overfull. Otherwise it will split the wood. Smoothing the bottom is done with a gooseneck

74

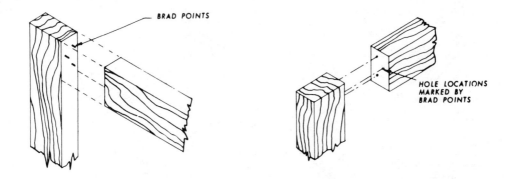

Fig. 3-18. Hole marking. Courtesy of Shopsmith, Inc.

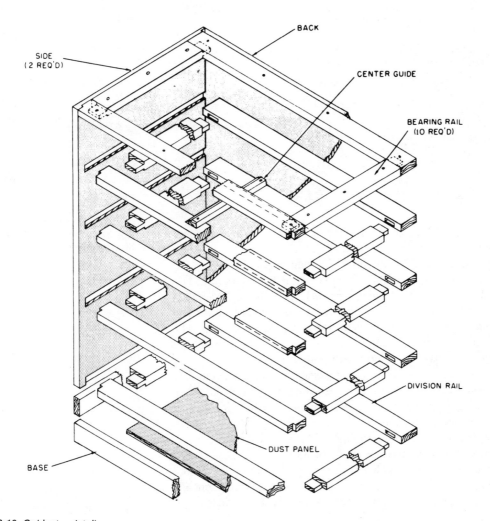

Fig. 3-19. Cabinetry details.

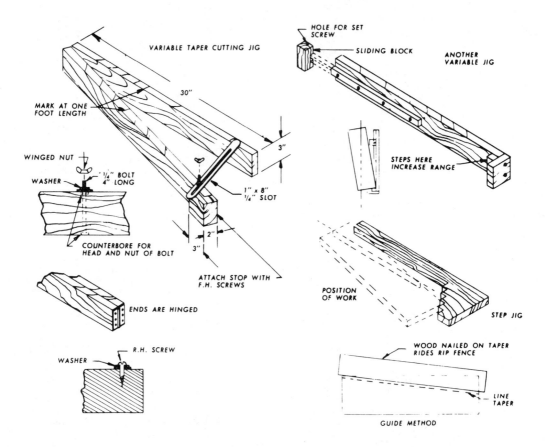

HOLE FOR SET SCREW

VARIABLE TAPER CUTTING JIG

SLIDING BLOCK

ANOTHER VARIABLE JIG

MARK AT ONE FOOT LENGTH

30"

3"

WINGED NUT

WASHER

1/4" BOLT 4" LONG

STEPS HERE INCREASE RANGE

1" x 8" 1/4" SLOT

COUNTERBORE FOR HEAD AND NUT OF BOLT

2"

3"

ATTACH STOP WITH F.H. SCREWS

POSITION OF WORK

ENDS ARE HINGED

STEP JIG

R.H. SCREW

WOOD NAILED ON TAPER RIDES RIP FENCE

WASHER

LINE TAPER

GUIDE METHOD

Fig. 3-20. Taper jigs. Courtesy of Shopsmith, Inc.

chisel. Smoothing the sides of the mortise is imperative for a proper fit overall.

The mortise and tenon should fit snugly. Because wood expands and contracts, too tight a fit will eventually cause splits. Too tight a fit also means all the glue, or nearly all of it, will be scraped off—providing very little holding power. This holds true for any wood joint where an adhesive is used to strengthen the grip. The space required is minimal, but must be there or you will not get enough grip to make the glue worth bothering with.

Other joint styles serve varied purposes, with some close to ideal for particular uses. Routers come in exceptionally handy when producing many of these joints. A good shaper also helps a great deal with many wood joints.

Router templates come in several forms. The simplest (and one of the most expensive because of its size and the required rigidity) is the template made for mortising door hinges into a door and jamb. This unit gives precise control over hinge placement and aids in setting hinge depth so that it is flush with the door edge.

Fig. 3-21. Right: Miter cuts. Courtesy of Shopsmith, Inc.

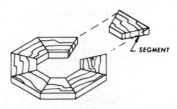

FOR OCTAGON (8 SIDES), SET
MITER GAUGE AT 67½°

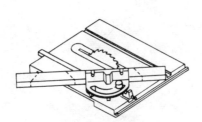

MITER GAUGE SET AT 45°
FOR CUTTING SEGMENTS
OF SQUARE (4 SIDES) AS
SHOWN BELOW

FOR HEXAGON (6 SIDES)
SET MITER GAUGE AT 60°

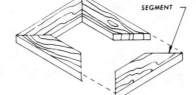

SEGMENT

4   3   2   1

CUT STOCK AS ABOVE FOR
BEST ECONOMY AND LEAST EFFORT

FOR BEVEL CUTTING, SET
TABLE AT ANGLE REQUIRED AND
MITER GAUGE AT 90°

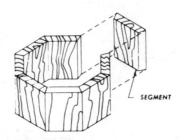

SEGMENT

FOR OCTAGON, SET FOR BEVEL OF 22½°

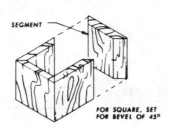

SEGMENT

FOR SQUARE, SET
FOR BEVEL OF 45°

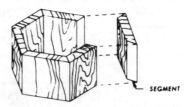

SEGMENT

FOR HEXAGON, SET FOR BEVEL OF 30°

77

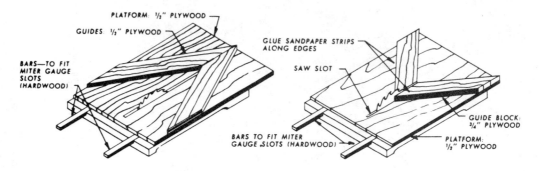

PLATFORM: ½" PLYWOOD

GUIDES: ½" PLYWOOD

BARS—TO FIT MITER GAUGE SLOTS (HARDWOOD)

GLUE SANDPAPER STRIPS ALONG EDGES

SAW SLOT

GUIDE BLOCK: ¾" PLYWOOD

PLATFORM: ½" PLYWOOD

BARS TO FIT MITER GAUGE SLOTS (HARDWOOD)

Fig. 3-22. Miter jigs. Courtesy of Shopsmith, Inc.

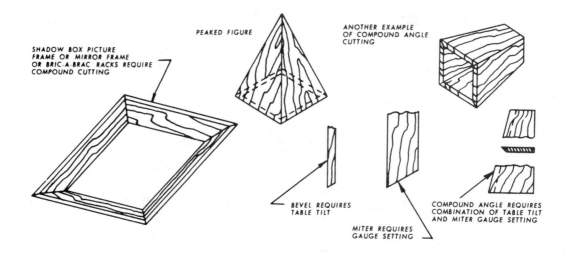

SHADOW BOX PICTURE FRAME OR MIRROR FRAME OR BRIC-A-BRAC RACKS REQUIRE COMPOUND CUTTING

PEAKED FIGURE

ANOTHER EXAMPLE OF COMPOUND ANGLE CUTTING

BEVEL REQUIRES TABLE TILT

MITER REQUIRES GAUGE SETTING

COMPOUND ANGLE REQUIRES COMBINATION OF TABLE TILT AND MITER GAUGE SETTING

WORK ANGLE	FOUR-SIDED FIGURE		SIX-SIDED FIGURE	
	table setting	miter-gauge setting	table setting	miter-gauge setting
15°	43¼	75½	29	81¾
30°	37¾	63½	26	74
45°	30	54¾	21	67¾
60°	21	49	14½	63½

Fig. 3-23. Bevel cuts. Courtesy of Shopsmith, Inc.

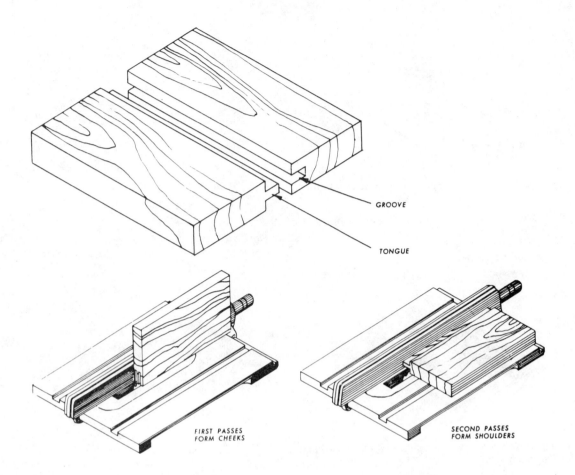

GROOVE

TONGUE

FIRST PASSES
FORM CHEEKS

SECOND PASSES
FORM SHOULDERS

Fig. 3-24. Cutting a tongue using a regular saw blade. Courtesy of Shopsmith, Inc.

Probably the simplest of all wood joints is the *butt joint*. With a butt joint, the ends of two boards are brought together and fastened to each other using nails, screws, glue, or something similar. This is not only the simplest woodworking joint, it is the weakest for almost any form of stress.

Fig. 3-25. Dadoing a tongue. Courtesy of Shopsmith, Inc.

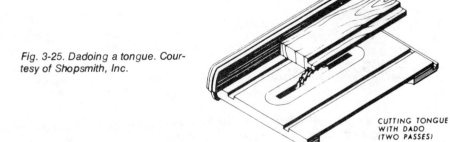

CUTTING TONGUE
WITH DADO
(TWO PASSES)

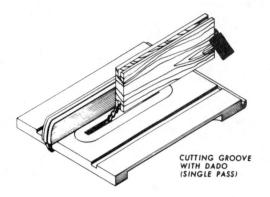

CUTTING GROOVE
WITH DADO
(SINGLE PASS)

Fig. 3-26. Dadoing a groove. Courtesy of Shopsmith, Inc.

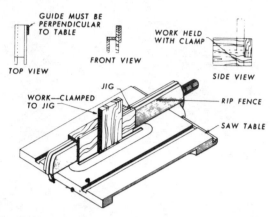

GUIDE MUST BE
PERPENDICULAR
TO TABLE

TOP VIEW

FRONT VIEW

WORK HELD
WITH CLAMP

SIDE VIEW

WORK—CLAMPED
TO JIG

JIG

RIP FENCE

SAW TABLE

Fig. 3-29. A tenoning jig is easily built. It must be made so that it slides easily on the fence. Courtesy of Shopsmith, Inc.

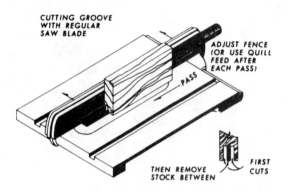

CUTTING GROOVE
WITH REGULAR
SAW BLADE

ADJUST FENCE
(OR USE QUILL
FEED AFTER
EACH PASS)

PASS

THEN REMOVE
STOCK BETWEEN

FIRST
CUTS

Fig. 3-27. Grooving with a regular saw blade. Courtesy of Shopsmith, Inc.

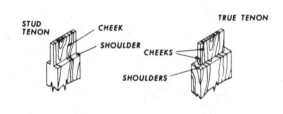

STUD
TENON

CHEEK

SHOULDER

TRUE TENON

CHEEKS

SHOULDERS

MAKING CHEEK CUT
ON TENON (TENONING
JIG MAY BE USED)

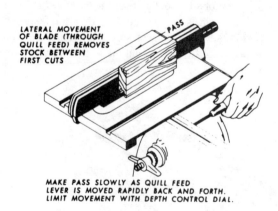

LATERAL MOVEMENT
OF BLADE (THROUGH
QUILL FEED) REMOVES
STOCK BETWEEN
FIRST CUTS

PASS

MAKE PASS SLOWLY AS QUILL FEED
LEVER IS MOVED RAPIDLY BACK AND FORTH.
LIMIT MOVEMENT WITH DEPTH CONTROL DIAL.

Fig. 3-28. Using the quill feed on the Shopsmith Mark V to remove material between saw kerfs when forming a groove. Courtesy of Shopsmith, Inc.

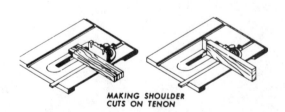

MAKING SHOULDER
CUTS ON TENON

Fig. 3-30. Making cheek and shoulder cuts with a regular saw blade. Courtesy of Shopsmith, Inc.

Some of the common woodworking joints that can be cut on the table saw, using a regular saw blade or a dado, are: 1—butt; 2—rabbet; 3—dado and groove; 4—splice lap (end); 5—middle lap; 6—end lap; 7—lapped miter; 8—dado and rabbet (box corners); 9—notch; 10—true tenon; 11—stud tenon; 12—slot (for stud tenon); 13—miter; 14—mitered bevel.

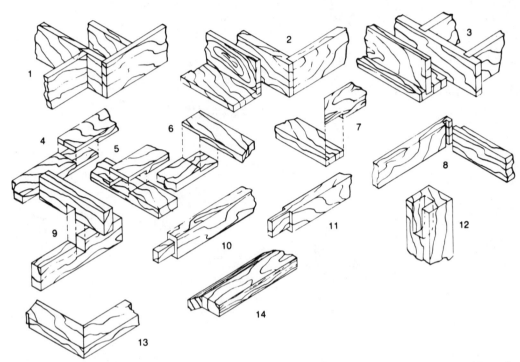

Fig. 3-31. Joints made with a table saw. Courtesy of Shopsmith, Inc.

In metalworking, a butt joint can be lengthened with a miter cut to increase the amount of metal facing. This metal joint is called a *scarf,* but in woodworking, as far as I know, it's simply called a miter or a mitered butt joint. Strength is increased marginally, but the greatest advantage of the mitered joint is that it makes things easier

Fig. 3-32. Beveled spline grooving. Courtesy of Shopsmith, Inc.

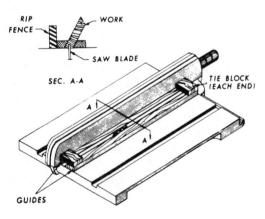

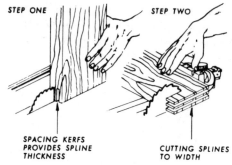

STEP ONE   STEP TWO

SPACING KERFS
PROVIDES SPLINE
THICKNESS

CUTTING SPLINES
TO WIDTH

Fig. 3-33. Cutting splines. Courtesy of Shopsmith, Inc.

GUIDE STRIP   PART ONE

FINGER LAP
JIG

DADO SET EXACTLY
FOR WIDTH OF CUT

PART ONE

PART TWO

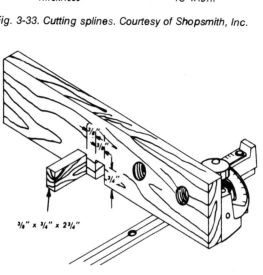

3/8" x 3/4" x 2 3/4"

Fig. 3-34. Finger lap jig. Courtesy of Shopsmith, Inc.

Fig. 3-35. Cutting finger laps. Courtesy of Shopsmith, Inc.

for you when you're using mechanical fasteners such as nails and screws.

Because there's more room to drive fasteners from one surface to the next, even when the miter is a straight-line joint (and not a 90-degree turn), the number and size of fasteners can often be increased (thus increasing the strength of the joint). Of course there's more surface for adhesives to work on so that type of fastening is also more effective.

Lap joints are quite simple. Two pieces of wood lap over each other. In proper lap joints, one or both pieces will be cut to fit the lap in the other. End laps make quite effective joints on storm and screen doors and in cabinetry. Center laps mean that the center of one board, or some point in from an end, is cut out to fit a lap cut in a crossing board. These are often found in the same types of work as are end laps. I use them when building certain types of cabinets, for forming the frame, and find that the ease with which the lap joint combines with the other styles (mitered lap joints at board ends provide a good locking action) makes for some degree of interest in a piece that might otherwise be exceptionally dull to make.

The *splice lap* is made at board ends. *Rabbet joints* are also easy to make with common woodworking tools. They can be cut with any table or radial saw using a standard blade (time-consuming and boring) or in a single pass with a dado blade. Essentially, a *rabbet* is no more than a cut that forms a lip on one board the same depth as the thickness of the board to fit into it.

It is not a great deal stronger than the butt joint, but does lend itself to ease of fastening, and can be used to hide plywood plies on board ends (in such a case, the plywood that is rabbeted would have the rabbet made deep enough to leave only a single-face ply). The *dado* and *groove* joint is also easily made with a standard blade.

To make such wide cuts with a standard blade requires enough passes for the saw kerf to widen to the point where the board to fit the dado or groove will slip in. Because most saw kerfs are under ⅛ of an inch, this can mean as many as 9, 10, or even more passes at a precisely set depth. A good dado blade can make the same cut in one or two passes. The dado joint is the short one; it usually passes across the grain of the board. The groove joint is identical except that it passes with the grain of the board, forming a longer slot.

The *spline and groove joint* is often used in place of tongue and groove joints. Spline grooves take splines, usually with glue, to reinforce joints. They reinforce butt joints particularly well and to the point many woodworkers prefer them when gluing up a wide board from narrow stock.

The board to be splined is simply run along the rip fence to make the groove. If the spline is wider than a single saw kerf, you'll need to run the board through with several different saw or board settings. The grain of the spline must run at an angle to that of the board it is fitting—that is, along the narrow width of the spline, not

Fig. 3-36. Finger lamps compared to butt, rabbet, and miter joints. Courtesy of Shopsmith, Inc.

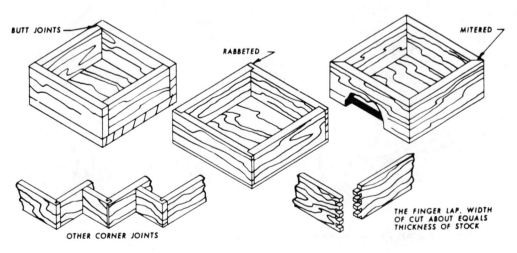

BUTT JOINTS

RABBETED

MITERED

OTHER CORNER JOINTS

THE FINGER LAP, WIDTH OF CUT ABOUT EQUALS THICKNESS OF STOCK

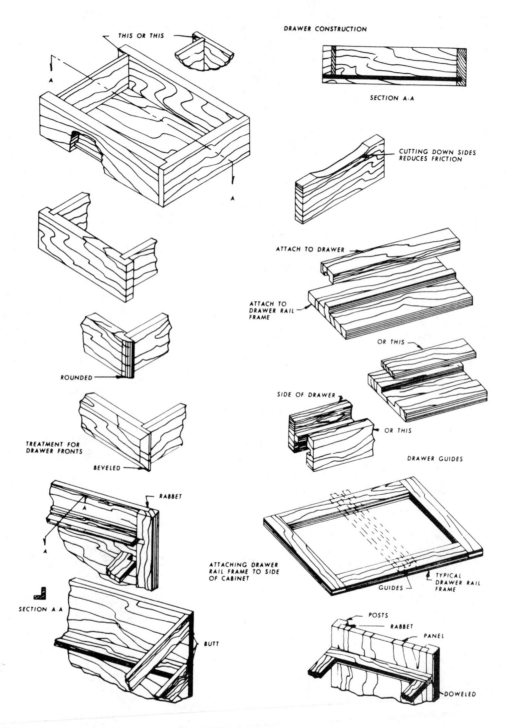

THIS OR THIS

DRAWER CONSTRUCTION

SECTION A-A

CUTTING DOWN SIDES
REDUCES FRICTION

ATTACH TO DRAWER

ATTACH TO
DRAWER RAIL
FRAME

ROUNDED

OR THIS

TREATMENT FOR
DRAWER FRONTS

SIDE OF DRAWER

OR THIS

BEVELED

DRAWER GUIDES

RABBET

SECTION A A

BUTT

ATTACHING DRAWER
RAIL FRAME TO SIDE
OF CABINET

GUIDES

TYPICAL
DRAWER RAIL
FRAME

POSTS

RABBET

PANEL

DOWELED

Fig. 3-37. Drawer construction details. Courtesy of Shopsmith, Inc.

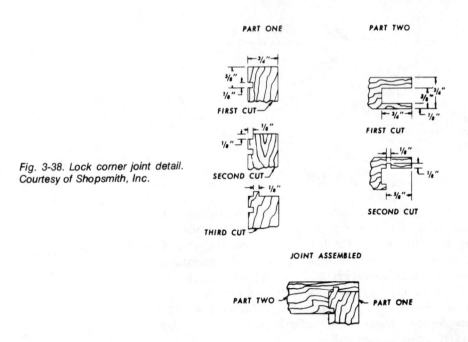

PART ONE

¾"

⅝"
⅛"

FIRST CUT

⅛"
⅛"

SECOND CUT

⅛"

THIRD CUT

PART TWO

¾"
⅛"
¾"

FIRST CUT

⅛"
⅛"

⅝"

SECOND CUT

Fig. 3-38. Lock corner joint detail.
Courtesy of Shopsmith, Inc.

JOINT ASSEMBLED

PART TWO   PART ONE

the spline length—if you are to get the proper strength. Depending on spline length, you'll size your stock and then re-saw the stock (using ¾-inch-thick stock) three times to produce three fingers approximately ⅛ of an inch thick. Then saw the board to remove the splines and repeat the process until you have sufficient splines. Splines are then sanded to exact length and thickness—a snug fit in the grooves.

Fig. 3-39. Wedge cutting jigs.
Courtesy of Shopsmith, Inc.

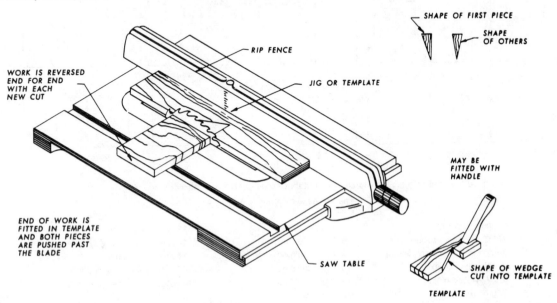

SHAPE OF FIRST PIECE

SHAPE OF OTHERS

RIP FENCE

JIG OR TEMPLATE

WORK IS REVERSED END FOR END WITH EACH NEW CUT

MAY BE FITTED WITH HANDLE

END OF WORK IS FITTED IN TEMPLATE AND BOTH PIECES ARE PUSHED PAST THE BLADE

SAW TABLE

SHAPE OF WEDGE CUT INTO TEMPLATE

TEMPLATE

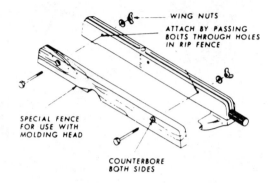

WING NUTS

ATTACH BY PASSING
BOLTS THROUGH HOLES
IN RIP FENCE

SPECIAL FENCE
FOR USE WITH
MOLDING HEAD

COUNTERBORE
BOTH SIDES

Fig. 3-40. An auxiliary fence for the molding head. Courtesy of Shopsmith, Inc.

Notching is another table saw chore where you can use a dado blade. Notching is basically a cut that moves the dado blade across the thickness of the stock and to about half its width. It is most often used to produce finger joints to make drawer assembly simple. Jigs can be made readily so the job goes quickly—with a dado blade. Finger width is generally equal to the width of the stock when used for drawers. For really small stock, heavy fingers look clumsy. Many people reduce finger size to ⅜ of an inch or so when stock is ¾ of an inch down to ⅜ of an inch.

Tongue and groove joints made on a table saw can also be made on a table saw or a radial saw using either a dado or standard blade. Again the dado is faster and easier to use, especially if you get one of the micrometer adjusting models that do not require you to remove

Fig. 3-41. Molding knife shapes. Courtesy of Shopsmith, Inc.

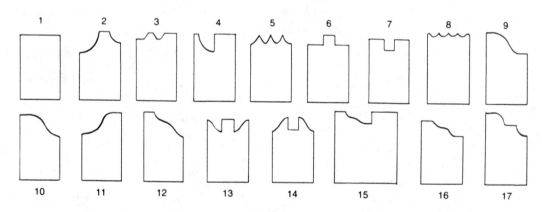

With a good assortment of knives you can turn out thousands of standard moldings and shapes of your own design. The knives shown above are:

1. blank cutter blade
2. ¼", ½" quarter round
3. glue joint, ½" stock and up
4. cabinet-door lip
5. "V"-flute molding
6. groove cutter
7. tongue cutter
8. four-bead molding
9. ogee molding
10. ogee molding
11. ogee molding
12. reverse ogee
13. panel-insert cutter
14. panel-insert (cope) cutter
15. 1⅜" sash cutter
16. 1⅜" sash cope cutter
17. bead-and-cover molding

Here are some cutter knives and examples
of forms that can be produced with them.
The knives are: 1—blank cutter blade;
2—¼", ½" quarter round; 3—bead and
cove molding; 4—ogee molding; 5—
"V"-flute molding; 6—ogee molding;
7—groove cutter.

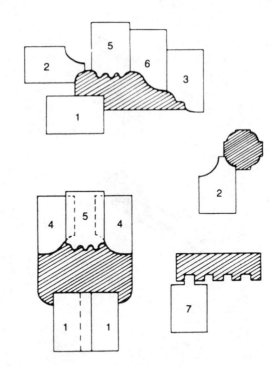

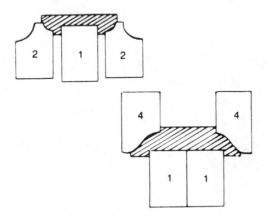

*Fig. 3-42. A few of the shapes you
can produce with molding knives.
Courtesy of Shopsmith, Inc.*

chipper blades to change cut sizes. The dado blade makes the
tongue cut possible in only two passes; a standard saw blade
requires four. The groove is cut in a single pass with the dado blade,
but might require four or more with the standard blade, depending
on the width of groove needed.

Mortises and doweled joints are best made with a drill press or a
drill (the drill press provides much greater accuracy, and can be
fitted with special mortising chisel drills such as those from
Shopsmith. *Dovetail joints* can be made by several tools, including a
router and a shaper (or shaper setup for a drill press).

Dovetail cutters are specially made for routers and shapers. The
high speeds and special steels give a very clean cut. Tenons use the
same cutters, but require two passes; the work is flipped over for
the second pass.

Dovetail slots or *grooves*, can also be cut with these special bits.
The tenon is formed to fit in the same manner as with the smaller,
finger-dovetail styles. the dovetail slot is produced in much the
same manner as is a dado, or groove, in whatever length is required.
This provides an excellent choice for rapid production when you
have a large number of drawers to make. It's also excellent for use
as a slot on shelves where there will be more than normal bending
stress (a standard slot or dado might lose its grip).

Dowel joints are simple and they are probably familiar to most

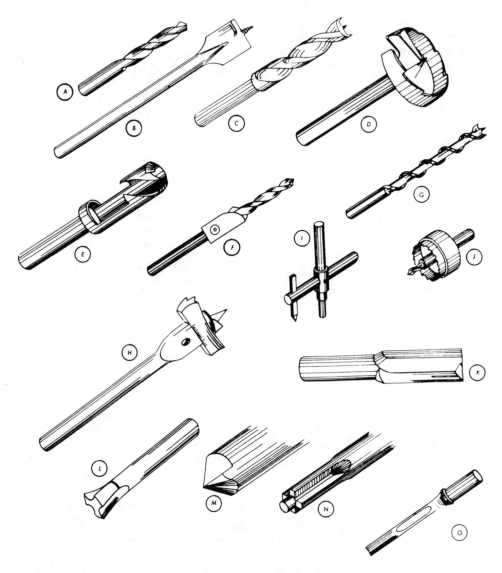

Shown above and listed below are most of the common tools used in the drill press:

A. twist drill

B. power auger bit

C. spur machine drill

D. multi-spur machine bit

E. plug cutter

F. adjustable countersink attachment

G. solid-center bit

H. expansive bit

I. fly cutter (hole cutter)

J. hole saw

K. router bit

L. dovetail cutter

M. countersink

N. counterbore (with center pilot)

O. hollow chisel

Fig. 3-43. Drill press accessories. Courtesy of Shopsmith, Inc.

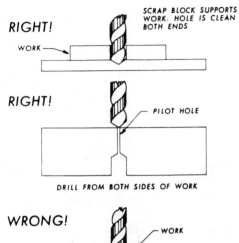

RIGHT!

WORK

SCRAP BLOCK SUPPORTS
WORK: HOLE IS CLEAN
BOTH ENDS

RIGHT!

PILOT HOLE

DRILL FROM BOTH SIDES OF WORK

WRONG!

WORK

DRILL BREAKS THROUGH
AND SPLINTERS WORK

Fig. 3-44. Scrap blocks prevent splitting when drill penetrates all the way. Courtesy of Shopsmith, Inc.

people who have ever seen a piece of furniture after it's fallen apart. A hole is drilled through two members and a dowel is inserted with glue. The members are clamped until the glue sets.

The dowel joint isn't that easily made, but it makes for speedy

Fig. 3-45. Screw hole styles. Courtesy of Shopsmith, Inc.

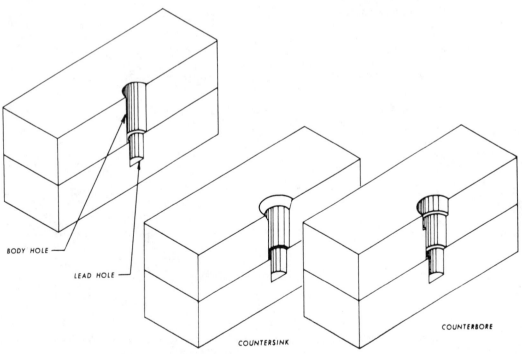

BODY HOLE

LEAD HOLE

COUNTERSINK

COUNTERBORE

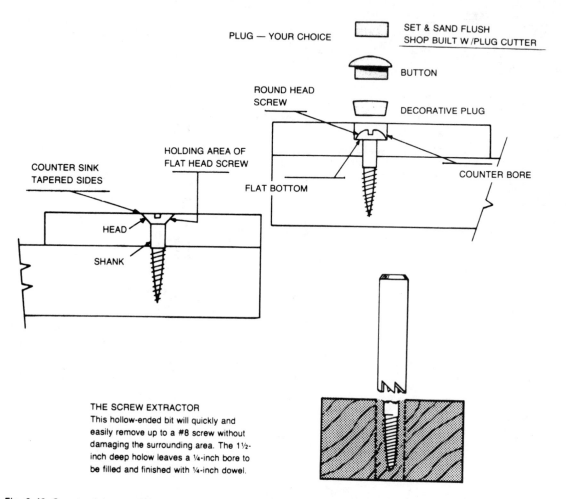

PLUG — YOUR CHOICE

SET & SAND FLUSH
SHOP BUILT W /PLUG CUTTER

BUTTON

ROUND HEAD
SCREW

DECORATIVE PLUG

HOLDING AREA OF
FLAT HEAD SCREW

COUNTER BORE

COUNTER SINK
TAPERED SIDES

FLAT BOTTOM

HEAD

SHANK

THE SCREW EXTRACTOR
This hollow-ended bit will quickly and
easily remove up to a #8 screw without
damaging the surrounding area. The 1½-
inch deep holow leaves a ¼-inch bore to
be filled and finished with ¼-inch dowel.

Fig. 3-46. Countersink, counterbore, plugs, extractor. Courtesy Shopsmith, Inc.

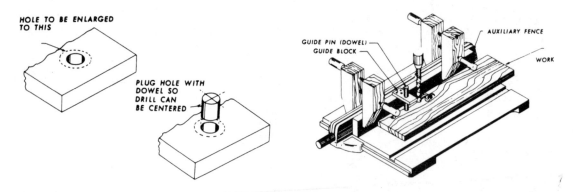

HOLE TO BE ENLARGED
TO THIS

PLUG HOLE WITH
DOWEL SO
DRILL CAN
BE CENTERED

GUIDE PIN (DOWEL)
GUIDE BLOCK

AUXILIARY FENCE

WORK

Fig. 3-47. One method for enlarging drill holes. Courtesy of Shopsmith, Inc.

Fig. 3-48. Screw hole center distance jig. Courtesy of Shopsmith, Inc.

90

SCREW GAUGE	BODY HOLE	LEAD HOLE	SCREW GAUGE IN INCHES
0	53	—	.060
1	49	—	.073
2	44	56*	.086
3	40	52*	.099
4	33	51*	.112
5	1/8	49*	.125
6	28	47	.138
7	24	46	.151
8	19	42	.164
9	15	41	.177
10	10	38	.190
11	5	37	.203
12	7/32	36	.216
14	D	31	.242
16	I	28	.268
18	19/64	23	.294

*In hardwoods only.

Fig. 3-49. Screw hole size chart. Courtesy of Shopsmith, Inc.

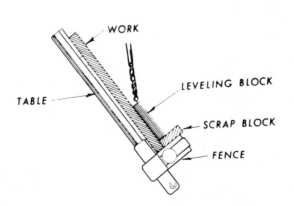

Fig. 3-50. Forming screw hole pockets. Courtesy of Shopsmith, Inc.

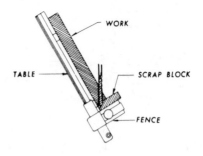

Fig. 3-51. Drilling extreme angles. Courtesy of Shopsmith, Inc.

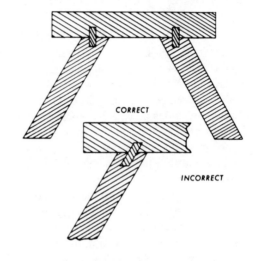

Fig. 3-52. Dowels must enter at a 90-degree angle to the surface. Courtesy of Shopsmith, Inc.

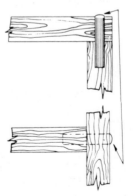

Fig. 3-53. Dowels may be rounded and left projected or sanded flush. Courtesy of Shopsmith, Inc.

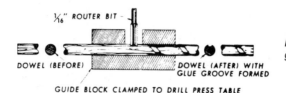

bracing of other types of wood joints or speedy assembly of a simple doweled butt joint with far greater strength than any plain butt joint.

Pegged joints are nothing more than doweled joints used to assemble boxes, drawers, and similar items. In combination with other joint styles, they can provide both strong and attractive joints.

JOINTS AND TOOLS

I use a Shopsmith Mark V for virtually all my stationary woodworking. Tool buying is something that can flat ruin many people, and I've reached the stage where my wife will nail my feet to the floor before she'll let me get near any place selling tools of any kind. She tried nailing my shoes down a few years ago but I went barefoot.

I'm going to assume that you've correctly set up your saw and made enough of the appropriate practice cuts, as follows, to be adept even if at a slow pace in using the various settings. See Figs. 3-63 through 3-80.

You should be familiar with the saw settings required to make crosscuts, rip cuts (with the grain), miter cuts (at an angle, a form of crosscut), cross-bevel cuts (with the table or blade at an angle), rip bevel cuts, compound miters (a miter cut made with the table or blade set at an angle), chamfer cuts (bevel cuts made only on board edges), a two-sided taper, a four-sided taper (these usually require jigs which can be bought or made), a compound rip-bevel, kerfing (individual or double, pass saw cuts at intervals along the board), bidirectional kerfing, and a two-pass rabbet (two crosscuts with the board in different positions).

These cuts can all be made with the basic saw blade, and no

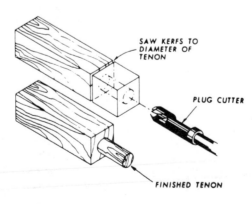

Fig. 3-55. Integrate dowels can readily be made. Courtesy of Shopsmith, Inc.

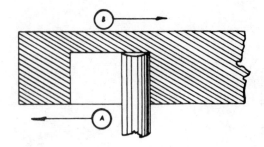

TWO FACTORS THAT MAY SPOIL A MORTISE

A. CHISEL TENDS TO LEAD OFF TOWARD THE CAVITY ALREADY FORMED.

B. WORK TENDS TO CREEP AWAY FROM THE CHISEL AS THE CUT IS BEING MADE.

ALWAYS MAKE END CUTS FIRST. OVERLAP CUTS SO THAT CHISEL ALWAYS MAKES A CUT THAT IS AT LEAST 3/4 SIZE.

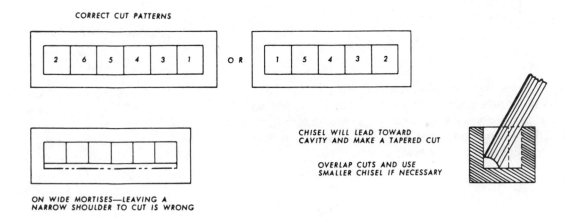

CORRECT CUT PATTERNS

| 2 | 6 | 5 | 4 | 3 | 1 |

OR

| 1 | 5 | 4 | 3 | 2 |

ON WIDE MORTISES—LEAVING A NARROW SHOULDER TO CUT IS WRONG

CHISEL WILL LEAD TOWARD CAVITY AND MAKE A TAPERED CUT

OVERLAP CUTS AND USE SMALLER CHISEL IF NECESSARY

Fig. 3-56. Using drill press mortising chisel. Courtesy of Shopsmith, Inc.

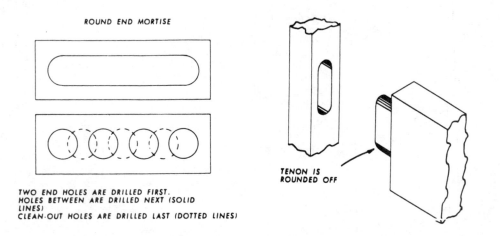

ROUND END MORTISE

TWO END HOLES ARE DRILLED FIRST. HOLES BETWEEN ARE DRILLED NEXT (SOLID LINES)
CLEAN-OUT HOLES ARE DRILLED LAST (DOTTED LINES)

TENON IS ROUNDED OFF

Fig. 3-57. Drilled mortise, with ends left rounded. Courtesy of Shopsmith, Inc.

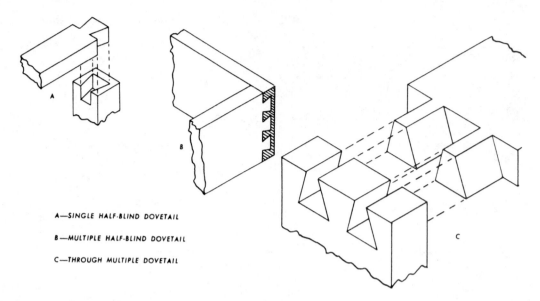

A—SINGLE HALF-BLIND DOVETAIL

B—MULTIPLE HALF-BLIND DOVETAIL

C—THROUGH MULTIPLE DOVETAIL

major accessories (including the taper cuts, but these are virtually always far neater and more accurate with a jig). You should gain some practice on the various cuts possible with both the *dado blade* and the *molding head*. It's at this point that the ease of making some joints will become really evident.

Fig. 3-58. Dovetail joints. Courtesy of Shopsmith, Inc.

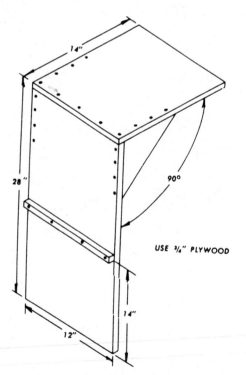

Fig. 3-59. Dovetail table. Courtesy of Shopsmith, Inc.

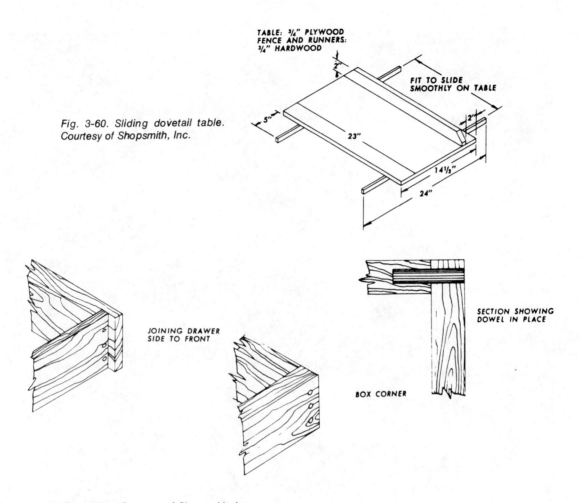

TABLE: 3/4" PLYWOOD
FENCE AND RUNNERS: 3/4" HARDWOOD

FIT TO SLIDE SMOOTHLY ON TABLE

2"

5"

23"

2"

14 1/2"

24"

Fig. 3-60. Sliding dovetail table. Courtesy of Shopsmith, Inc.

JOINING DRAWER SIDE TO FRONT

SECTION SHOWING DOWEL IN PLACE

BOX CORNER

Fig. 3-61. Doweled joints. Courtesy of Shopsmith, Inc.

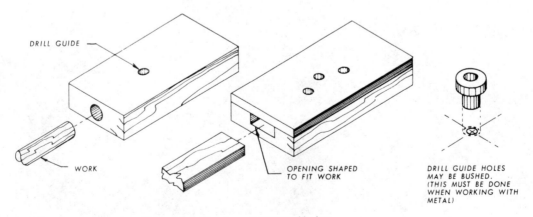

DRILL GUIDE

WORK

OPENING SHAPED TO FIT WORK

DRILL GUIDE HOLES MAY BE BUSHED. (THIS MUST BE DONE WHEN WORKING WITH METAL)

Fig. 3-62. Duplicate piece dowel hole pegs. Courtesy of Shopsmith, Inc.

95

Fig. 3-63. Using the horizontal boring feature to drill dowel holes. Courtesy of Shopsmith, Inc.

Fig. 3-64. The Shopsmith Mark V offers an almost unique horizontal boring feature to ease alignment of dowel holes after drilling. Courtesy of Shopsmith, Inc.

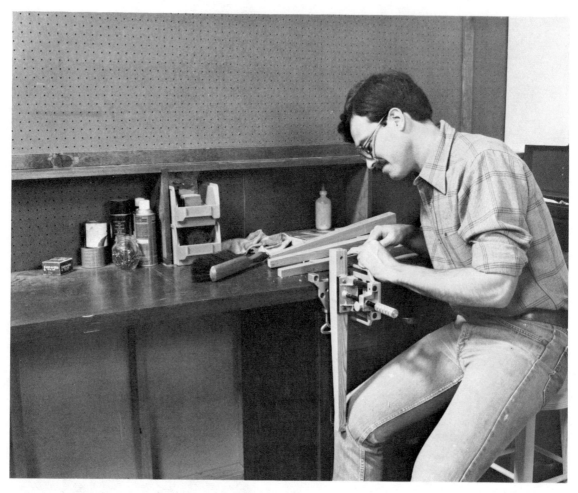

The dado, of course, is made with the dado blade. Most dado blades require only a single pass, but very wide dadoes might require two passes. The groove is also made in a single pass, most of the time, using a dado blade along the grain of the board. In those two cuts you've got the basis for all kinds of shelf-building activities.

Hollowing cuts can also be made with the dado blade and, of course, dadoes can be stopped at any point required so that the board edge doesn't show at the face of the work. If you use a stop dado, the board fitting into it will need to be cut in sort of a boat shape to fit the curve of the saw where the dado was stopped (or you can cut it square and chisel out the dado to fit).

Rabbets come in various styles and shapes. They can be run across the grain on the board or with the grain as needed to fit a particular joint style. Notching is done with the dado head. This is tongue-and-groove cutting (with the proper cutters, tongues can also be made on the molding head, as can grooves). Stud tenons are

Fig. 3-65. The final assembly of an attractive project is the aim of wood joint making. Courtesy of Shopsmith, Inc.

Fig. 3-66. Clock case assembly requires careful joinery, proper glues, and correct clamping. Courtesy of Shopsmith, Inc.

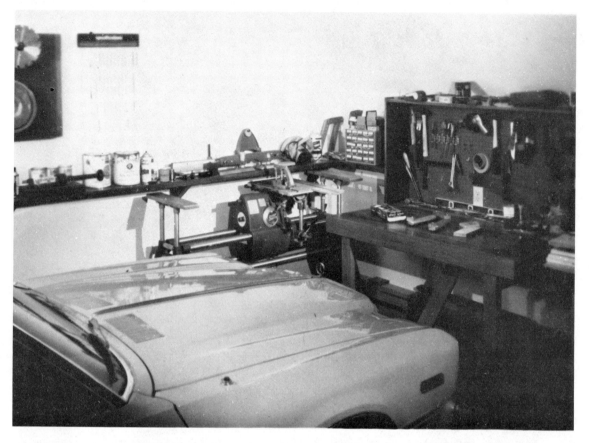

easily made on the dado head (as are slots, both stopped and through).

True tenons have shoulders on all sides, and might or might not pass all the way through another member. *Coving cuts* are all made easily with the standard saw blade. They can be widely varied by adding passes from the dado blade to give a flatter surface. The blade requires a great deal of adjustment for this sort of cutting, as it does with surface cuts made with a molding head.

Fingerlap joints are also made with the dado head, but dovetail joints are most frequently made with a shaper or a router, and not with any accessory normally used on a table saw of any kind. Special grooves are readily made with the dado blade in almost any design you choose to produce.

Crosscuts, no matter the type blade used, are most easily made with the miter gauge and safety grip down. Free pieces are never picked off the table while the blade is still running. Miter gauge extensions can be built to handle especially large or oddly shaped work. When crosscutting, you should always use the miter gauge on your saw. Your hands should never approach the blade.

Any photos you see in this book of a full depth cut not using a safety guard over the blade are for demonstration purposes only, for clarity of photographs.

Ripping is the action of reducing the width of stock, and it's with this cut that such items as pushers and various guides that are held in your hands and used on the stock, guiding it through the blade, are most useful. It's usually with rip cuts that your hands must come closest to that sharp, spinning blade. Most experts recommend some sort of pusher stick whenever your hand is likely to come within about 4 inches of the blade.

Pusher sticks are simply shaped more or less to fit your hand. One end is rounded a bit so that it doesn't gouge and does give decent control. The other end is notched to fit the board being cut. Pusher blocks, or pusher-holddowns as they're also called, are used in grooving, edge-rabbeting, jointing, and various shaping operations as well as when ripping stock. They need consist of only one

Fig. 3-68. Shopsmith Mark V.

101

Fig. 3-69. Horizontal boring with the Mark V. Courtesy of Shopsmith, Inc.

handle, if you like, though two will provide better control. The Shopsmith version is actually shorter, with a single handle, so that you can use one or two as the work demands.

Mitering cuts require setting the miter gauge, and are most often crosscuts. Nevertheless, you might want to produce some rather long, slightly angled, cuts at one time or another.

Miter cuts are often combined with *bevel cuts* to make joints for frames, planters, and other types of boxes. Work is held lightly against the miter gauge head and the joint is cut. Use a miter gauge holddown whenever possible because the hold-down counteracts any tendency of the work piece to creep while being cut.

Both miter and bevel cuts require some care in the making. More care and less speed are required when the two cuts are combined to make a compound cut. First, the blade must have

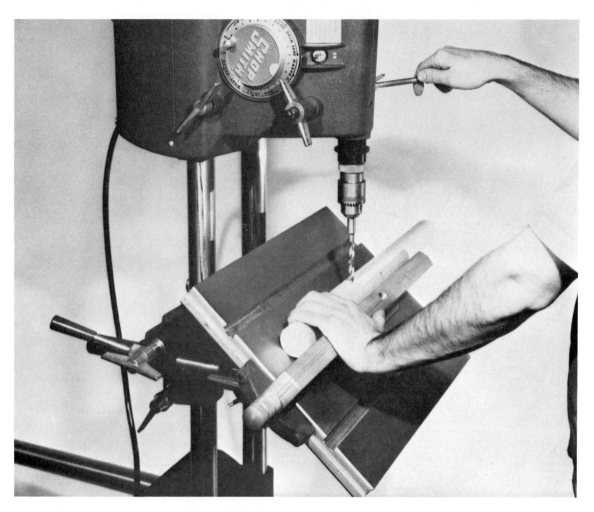

correctly set teeth and must be sharp. Next, you must get your table
(or, if the saw is of the type with a tilting arbor, the blade) angle just
right. The wood used must be as close to perfect as possible.
Warped or cupped wood will not produce decent angles.

Fig. 3-71. Using the Mark V as a vertical drill press. Courtesy of Shopsmith, Inc.

correctly set teeth and must be sharp. Next, you must get your table
(or, if the saw is of the type with a tilting arbor, the blade) angle just
right. The wood used must be as close to perfect as possible.
Warped or cupped wood will not produce decent angles.

The work cannot be allowed to creep as the cuts are made.
Passes must be made at a steady, slow pace. Just jamming the work
through the blade causes, among other things, poor control and loss
of accuracy. Take it easy and all will be well.

Because miter cuts must match perfectly when you make a
number of them, it's a good idea to buy a device known as a
miter-gauge stop rod. This rod can be set to a length needed. The
repeat cuts for a number of pieces will be as close to identical as
you're about to get. The gauges are set off the first miter made so it
must be dead accurate.

Matching left and right hand miters can be a problem. Using

Fig. 3-72. Right: The lathe in use. Courtesy of Shopsmith, Inc.

104

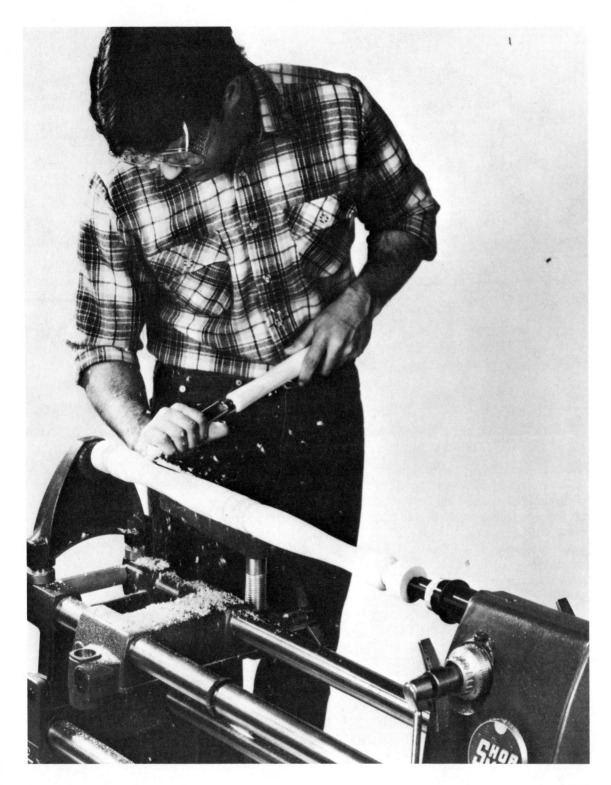

Fig. 3-73. Jointer use with the Mark V. Courtesy of Shopsmith, Inc.

sliding tables made for the purpose, you can miter any piece, right or left, and have it match another. The only real problem with such sliding tables is that they limit you to a single angle. If the guides are set to produce dead-on, 45-degree miters, then that's what you're going to get each and every time. You'd need a board for different angles, one for each angle needed, if you change off much.

Beveling calls for a table or arbor tilt and the resulting cut might cause some confusion. *Cross bevel* cuts, when joined, form miter joints. I've never heard of a bevel joint. It's simplest to call all cuts, needing the use of the miter gauge, miter cuts. Bevel cuts are made with the miter gauge set at 90 degrees. The work needs as much support as possible, with a very slow feed into the blade.

Compound-angle cuts require the miter gauge setting for the appropriate angle, plus the table or blade, tilted to the correct angle all at the same time. Such cuts are needed when you decide to make tops for fenceposts and other such items. Compound-angle cuts are among the most difficult table saw operations if you want perfect work. They require a good deal of practice and careful checking of settings so that you're certain of each setting before you feed the good stock through the blade. It often makes sense to feed scrap stock part way through to check angles.

On the left side of the blade, the angle will be limited because a few settings will force the miter-gauge head into the cut line. You should always make a test pass with the power off to see if this might happen.

Compound miter joints are easier to assemble, and far stronger, if they're combined with a spline set into grooves. To do so, without changing blade settings, move the rip fence in close to

Fig. 3-74. While a planer seems like an esoteric expensive machine for a home shop, a careful check of wood prices will show where savings can readily be found. Courtesy of Shopsmith, Inc.

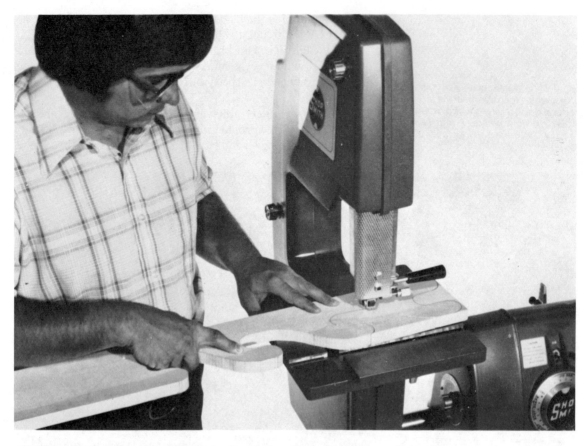

the blade and lock it at the spot where the blade will cut a groove dead center in the stock. On the Shopsmith, use the quill feed to make the precise blade setting. Set the table height so your groove will be half the width of the spline used (again, plus a very small fraction of an inch so that there's room for glue to move).

Hold the stock on edge and make your pass through the compound miter cut end. When making such passes, always hold similar faces of the work to the rip fence. If you make the groove on one piece by cutting with the inside face of the figure to be formed against the fence, then all the pieces must be cut with the inside face against the fence.

Dado and groove cuts are almost always wider than the saw kerf of a standard blade. A dado-blade assembly or micro-adjustable dado blade is used. A basic dado saw blade set will consist of two saw blades, always smaller than the full-size, table-saw blade, and several chippers, depending on the dado width required. The chippers are added or removed as required. You must use a special table insert to accommodate the extra width of the dado unit.

Dadoes are always made blind, or less than the thickness of the

Fig. 3-75. The Mark V bandsaw. Courtesy of Shopsmith, Inc.

Fig. 3-76. Right: Belt sander. Courtesy of Shopsmith, Inc.

stock being cut. If the cut must be deep, in thick stock, you'll find it far easier to use more than one pass to make the entire cut. Beyond those tips, and the need for a slow and steady feed into the dado blade, dado and groove cuts need no other special considerations.

Groove cuts are made in the same manner as are dadoes, but they are made with the grain direction instead of against it. With the long groove sometimes forcing the work to rise off the dado blade, it's a good idea to use pusher sticks or blocks, or to use a holddown attached to the rip fence to keep the work in place as it's fed into the blade.

You control both dado and groove width by adjustments to the rip fence for those times when more than a single pass is required. You can make grooves as wide or as narrow as required, even if your dado blade cuts no wider than ½ inch. This is the case only on smaller saws and with less expensive dado head sets. Dado heads are not the most expensive item you'll ever buy anyway, and even the micro-adjustable types that require no disassembly before changes in cut width are quite reasonable.

Spline grooves in noncompound cut stock are easily made, but require care so the splines are straight along the stock when a long groove is needed. If the work is simply beveled, a set of guides can be easily made so the cuts are at the proper angle to the bevel angle. The use of a push or spring stick keeps your hand well out of harm's way otherwise. If splines are used only to reinforce short joints, such as in picture frames and door and window moldings, the work is called *feathering*. A jig that slides easily over the rip fence makes cutting the grooves for feathering a great deal easier.

Notching cuts are dado cuts across the stock thickness and to half its width. When used for finger joints, the notches might well not extend half the stock width. Finger joints are very simple and quite easy to make. You control the notch width and depth with the saw blade. The crucial dimension is the depth of the notch. This must be identical all along and on both pieces. Once that is set correctly, all you need do is make the notches in both pieces, use the same saw setting, and be careful to hit your marks.

The finger-lap joint is used in drawers (although the dovetail is stronger) because it is easy and quick to make, and because it provides a lot of surface for gluing. Finger-lap joints are quite attractive. They are often used when the joint will be visible.

You can quickly build a jig to allow you to cut both halves of a finger-lap joint at the same time, and the jig is good in another way. It can stand repeated use and doesn't limit the width of the work. You can use it to make drawers or boxes that are shallow or deep.

The jig is simple to make, but the measurements used are crucial if you want your joints to fit together well. Make certain the dado is set for a cut exactly ⅜ of an inch wide, using a ¾-inch table

Fig. 3-77. Right: The jigsaw: this is an under-used tool, with a saber saw adaptability that allows easy cutting of the insides of circles without removing or breaking the blade. Courtesy of Shopsmith, Inc.

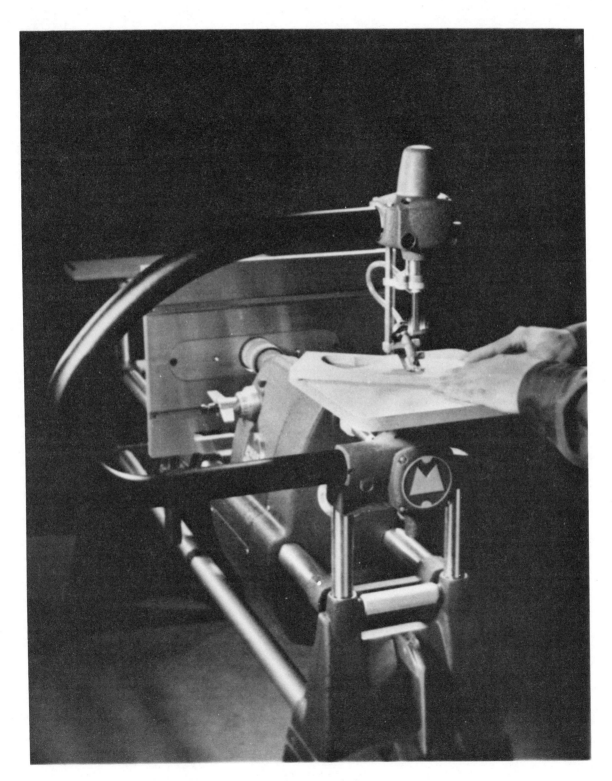

height (if the stock is to be ¾ of an inch thick). When you get ready to cut a notch in the miter-gauge extension, make sure the extension will fit snugly in the miter-gauge slots and the locknuts are well tightened before you make your feed. The second notch is made exactly ⅜ of an inch past the first one, and a guide block exactly ⅜ inch by ¾ inch by 2¾ inches long is glued in place in the second notch.

Make the first cut, in one piece only, using a guide strip. Make sure it's ⅜ of an inch thick. The L-shaped cut formed is then butted against the guide block and the mating part can be slipped into place. Any remaining cuts are then made. Spacing is automatic when you fit the preceding cuts over the guide block.

This type of jig serves well only when the same thickness stock is continuously used, but you can add variability to it by slotting holes for bolts holding it to the miter gauge. Use great care to make sure the jig is set up properly for each job. Gauge cuts by holding their inside faces firmly against the guide pin. Otherwise, you will have a sloppy fit every time.

Tenons and slots are differing cuts. The tenon is a projection, from the solid stock, of a piece of material that fits into a cavity, known as a mortise, in another piece. The mortise must be made on a drill press or by hand.

Tenons can be cut using a miter-gauge stop rod to make sure all four shoulders—on a true tenon—are equal or by using the tenoning jig shown in Fig. 3-29. Tenons may also be made with jointers and handsaws with excellent results. To make a tenoning jig for the table saw, you produce a jig that will ride the rip fence (slide along it readily and smoothly). The stock being tenoned is then clamped to the jig and fed through the blade.

If you want to cut the cheek and shoulder of the tenon at the same time, you will need a dado head. Width of the tenon cut is adjusted by moving the rip fence. If a standard saw blade is used, you use the jig to make the cheek cuts first and then make the shoulder cuts, using a miter-gauge, with the blade projecting just far enough above the table to meet the cheek cuts. There's a distinct similarity here to cutting rabbets, only on all four sides of the true tenon or two sides of the stud tenon.

You can use a slot for the stud tenon mate and cut that slot with the dado blade, using a stop block to set the length of the cut. The dado will leave a radius at the end of the cut. You can either use a chisel to square that off to fit the edge of the tenon or you can run the groove long enough so the radius doesn't interfere with the fit.

Tongue and groove joints on a table saw require a number of passes with a standard saw blade. When a standard or single saw blade is used, the tongue cut requires four passes to form the tongue. Using a Shopsmith saw, you can remove the material from

Fig. 3-78. Right: Making a compound rip-bevel cut. Courtesy of Shopsmith, Inc.

the groove more quickly than with a standard saw. The quill feed can be used to give lateral movement to remove material between saw kerfs. Otherwise you simply keep making kerfs until you've got the proper groove width.

With a dado blade, the tongue requires two passes and the groove requires only one. The tongue is cut, shoulder and cheek, in a single pass and then flipped to cut the other side. Only a shaper or molder can do it more quickly in a home shop.

Tongues, as with other tenons, are made just a fraction shorter than the groove is deep so you'll allow room for excess glue. If the joint's not to be glued, the extra space is not needed. Even then a fraction on a long joint is worthwhile so you can be sure no high spots in the groove will keep the tongue from bedding all the way in.

Wedges for wood joints are often used to tighten things up without glue, and permit easier later disassembly. Wedges can be cut or bought, but the ones you buy are almost always made of exceptionally soft material that might not match the exterior appearance you prefer.

Wedges are often used with tenons. There's a technique for easy cutting of the wedges using a jig that takes only a few minutes to make. You can cut them freehand but the jig produces far more uniform results. Cut the shape into a guide block—whatever shape you require—making sure the guide block is at least five times as long as the wedge. Also make sure the wedge to the inside tooth edge of the blade is the right size, with the guide set against the rip fence. Make the cut with the stock set so the grain runs the long way in the wedge. Wedges with the grain across them will snap off as they're driven.

When the time comes to use the wedges, simply split the end of the piece to be wedged and drive the wedge into place. Use a soft-faced mallet or hammer. As you make wedges, the material being cut is reversed end for end to keep shapes uniform.

The molding head, for use on a table saw, requires an even wider insert than does the dado blade, but you produce exactly what its name implies—all sorts of molding. Various cutter shapes are available for the various types of decorative and useful moldings. Specific molding heads allow you to cut several kinds of joints in fewer passes than is required with other tools, including a dado head.

The molding head is moderately expensive if you select many cutters to go with it, but in general it is one of the more worthwhile purchases you can make.

Molding cuts that must be deep are made in more than a single feed so the surface will not be marred with blade chatter or burn marks. As with most cutting tools, the best and smoothest cuts will come when you're working with the wood grain instead of against it.

Fig. 3-79. Right: Drill press. Courtesy of Shopsmith, Inc.

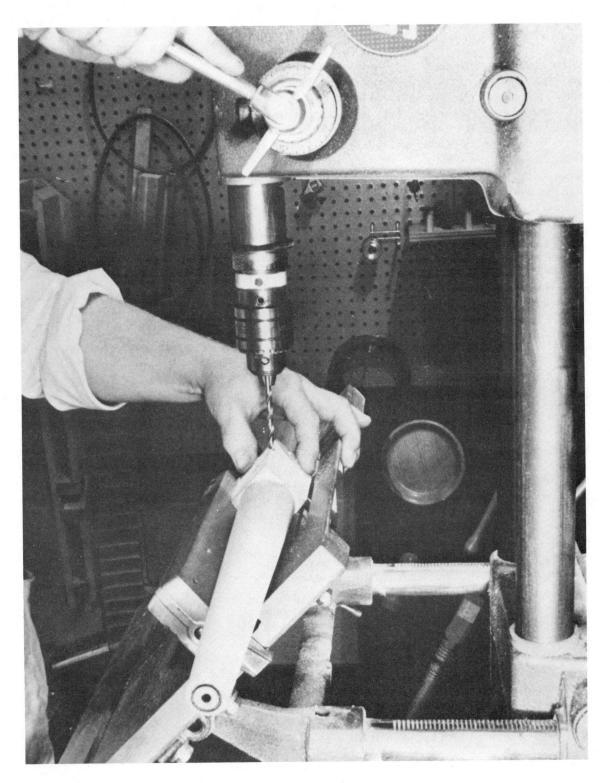

Use the miter gauge to feed crosscuts but feed them very slowly. If you're working all four edges of a piece of material, make cross-grain cuts first. Cuts along the grain will clean up the work and remove any splintering those might cause. Make cross-grain cuts, even in this case, very slowly to reduce problems.

Remember that the profiles of molding cutters shown here and those you see elsewhere are not all the cuts you can make with these knives. A bit of practice and thought will show the kind of results you can get making multiple passes after moving the work piece, or in not setting the cutter head so the knives cut all the way into the work, thus using only a portion of the profile.

As an example, shallow glue joints for stock just from ½ inch and up can be repeated as needed along the work's edge by simply turning the workpiece over. Groove cutters are usually limited to one size, but for really thick stock you can pair tongues and grooves, one above the other, in the edges of the work.

The tongue cutter can be worked in the same manner. It's at this point the molding cutter head and knives start showing their simple usefulness. Tongues are cut with a single pass along the board edge. Not two passes, as with a dado blade, and not four as with a regular saw blade. One pass.

THE JOINTER

Jointers are relatively simple tools, with a split table—split crosswise—running over a knife head, and also with a fence to guide the work. The basic use of the jointer is simply to get a square, smooth edge on any workpiece before proceeding to other operations.

Rabbeting with a jointer is a fairly fast process. Rabbets up to the maximum width of the jointer knife cut can be made in a single pass (assuming a shallow rabbet). Keep the cuts to ⅛ inch for good results or less for best results. Always use a pusher or pusher-holddown tool when rabbeting across the grain.

The jointer's rabbeting feature can also produce tenons by rabbeting on both sides of an edge, to form a stud tenon or tongue (on long stock), or by rabetting all four sides to form a full tenon.

From this point you can get inventive. Tenons and tongues can be beveled or chamfered. This is seldom necessary (or even wise), and the jointer goes on to make tapered shapes, octagonal shapes and to do many other jobs. For wood joinery, the jointer is used to true and smooth board edges (for edge-to-edge work, whether for butt joints or doweled joints, this procedure is almost essential to a good joint) and to form rabbets for lap joints and grooves and tenons.

THE DRILL PRESS

For any through work, drilling is best done from both sides of the

Fig. 3-80. Right: Drilling at a mild angle. Courtesy of Shopsmith, Inc.

work surface to prevent splintering. You can also clamp a piece of waste wood to the backside of a hole that will be hard to reach or to save turning the work over and lining the bit up. I like this method and use it most often, but I also either use a contrasting color wood or slip a sheet of paper between the pieces. Doing it that way keeps you from wasting time drilling all the way through the waste piece. As soon as the drilled out waste shows a color change or bits of paper you can stop drilling.

Screw holes are required in great variety for many projects. One useful item is the V block. This is only needed for drilling holes in round stock, but it can be used for various-sized round stock as well as stock of almost any material. They can be made fancy or plain or long or short stock. Long V blocks should have a clamp block for at least one end of the stock being drilled so you can be sure things are properly aligned and won't slip as the bit enters the work.

Most all drill presses will offer you a stroke of about 4 inches. That is the length of the most common drill-bit shank extending from the chuck tip. With longer drill bits, you move the stock closer to the bit tip as you drill. For horizontal boring, move the power head closer to the stock and drill again. A double operation would give you an 8-inch-deep hole (with an 8-inch-long bit, which I doubt you'll find, as most are 10, 12, or 18 inches long).

When I'm drilling for lamp bases, I use an 18-inch bit (the standard length for most lamps, though by careful centering you can double that by drilling in from both ends). The standard screw thread lamp rod to carry the cord, attach the fixtures, etc., requires a ⅜-inch-diameter hole. You will need to make a total of five operations if the material to be drilled requires the full length.

Drilling at angles in stock is often required for attaching table legs, shelf supports, and so on. In most cases, where the angles are not really extreme, you can simply tilt the table and drill right on into the stock. This works well on simple angles for items such as chair legs that tilt in only one direction. If the angle is compound, and that chair leg must tilt in two directions, use the miter gauge and a clamped-on piece of wood to form a V block on the table. Unequal compound angles, with the chair leg tilting in two directions as well as tilting different ways, requires the use of the V and a tilt of the table as well.

If you must also fulfill a need for screw-hole pockets, some special steps are needed. Screw-hole pockets are often needed to attach rails or frames to the underside of tables or cabinets or desks. Most often, a sharp angle is needed for the pocket, and this angle will kick the bit tip off its mark almost all the time. A jig is needed to allow you to drill the hole. Actually, a scrap block clamped to the work surface is most often used. The block is set so that the workpiece is level with the hole starting in the scrap block (Fig. 3-51) before it hits the work. This is especially useful when the tip angle is so extreme the tip will not meet the work until after the side of the drill touches. The leveling is actually just in relation to the drill bit.

Drill press joints are made with relative ease and speed by using hollow chisels (with drill bit centers) to make mortises. The major joints are mortise and tenon, blind mortise and tenon, twin mortise and tenon, key joints (a mortise and tenon with two sides open, or made as any other mortise and tenon), and three-way joints (two mortised sections setting on a full tenon), a haunched mortise, and a pegged tenon.

The *mortising chisel* is chucked into the drill press, using a special holder, after which you make certain there is a gap of at least 1/32 of an inch between the cutting bit and the edge of the hollow chisel. The side of the chisel is kept at right angles to the drill press fence. Shopsmith makes a special mortising holddown—suitable for about two dozen other uses as well—that is set into a hole in the fence and serves to hold the work in place. Without a holddown, the work tends to lift as the quill on the powerhead is withdrawn, lifting the chisel from the work.

Cuts for a mortise are best made with the end cuts made first. After that you make the cuts to clean out interior stock. Make sure that at least ¾ of the width of the chisel comes into play. You might not be able to use that width on your last cut, but use as much as possible. If the mortise is to be a really wide one, use smaller chisels. That sounds strange but leaving a narrow shoulder means a messy mortise. Leaving a wider working area for the second pass is a far better idea. This provides a square mortise about as rapidly as you'll be able to make one with any tool I know of today. It also will provide a near clean blind mortise bottom.

Mortises can also be cut with a drill bit. Drill the end holes first, and then drill intermediate holes to clean out the intermediate stock. Set the quill in its down position, finally, and move the stock back and forth along the drill bit to smooth the edges. The mortise so made will have rounded ends, and the tenon can either be cut to fit the radius of the ends or you can use a hand chisel to square the corners.

Dovetail joints are made with the drill press and a special dovetail bit. The strength of the joint comes because the dovetail is shaped to resist stress in all but the direction from which its tenons are inserted. There are other joints that offer this feature, but they get so complex they're harder to make than the dovetail.

A special dovetailing table can add speed and aid accuracy when making this joint becomes something you do with any frequency at all. A sliding table works exceptionally well in producing joints at the ends of long stock. It makes cutting slots in wide stock far simpler and more accurate.

Doweling can be easy or complex, depending on joint needs. It can be an almost rough-and-ready method of joint repair or it can form attractive joints for many uses.

Dowels work well in a number of joint styles, providing good support, and, usually, great ease of assembly (ease of assembly is almost always a factor of how carefully you measure, mark, and drill or cut a piece of work). Note that dowels enter at right angles to surfaces and not at the same angle as the piece being mated may enter.

Two dowels are far better in even a small joint than is a single dowel. The mated pieces can always twist around a single dowel. Adding a second makes twisting close to impossible. Miter joints are often dowelled to add strength. The drilling is best done with the miter gauge used to hold the work in position and the work firmly clamped to the table.

Dowel pegs can be used as design detailing, either extended and rounded off, or sanded flush and covered with paint. For a really startling look in some pieces, dowels can be made of contrasting wood color with a clear finish used.

Grooves or flutes are needed in dowels. It seldom pays to use a plain-surface dowel if the joint is to be glued. There is simply no place for excess glue to go. Many companies sell fluted dowels, usually with straight-line flutes. I keep bags of these on hand because the uses are so many, with sizes ranging from ¼ inch by 2 inches on up to ½ inch by 3 inches. They're reasonably inexpensive. If you prefer, a simple jig will allow you to cut curved flutes (considered slightly better in holding power than dowels with straight flutes) for yourself.

Chapter 4

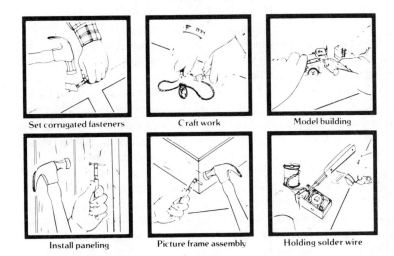

Set corrugated fasteners   Craft work   Model building

Install paneling   Picture frame assembly   Holding solder wire

Adhesives and Glues

The Egyptians used adhesives and glues thousands of years ago to fasten veneers and inlays to other materials, and the Greeks and Romans used cements and mosaics to apply ceramic tiles. We still find that hide, hooves, and horns are boiled together in a vat, as it was in the early days, to form hide glue. Such glue might come to you dry for mixing, for use over heat, with water, or it might come ready-mixed.

I've got several plastic bottles of Franklin's Hide Glue around, and it occasionally gets used when I need a long assembly time (clamping time can be two dozen hours). You can take a couple hours to set things up. The stuff is ideal for complex assemblies with good strength. The resulting joint is stronger than the wood it holds together. It also fills gaps a bit better than some other glues we use with greater frequency these days, and is not affected by most solvents.

While hide glue works only on porous materials—it's at its best on wood and leather—modern glues and adhesives will bond almost anything to anything else. It was really the World War I era that saw the development of casein glues, produced from milk, and nitrocellulose glues. Nitrocellulose is the same material the British call gun cotton and it is a main factor in the making of gunpowder.

During the 1930s, plastic resin glue—urea resin and resorcinol resin—came on the market. The resorcinol glue really was the first truly and totally waterproof glue. Many new adhesives and glues were developed during World War II, but most didn't come to the marketplace until well after the hostilities stopped. We didn't get to see the epoxies and acrylonitriles until the early 1950s.

Since the 1950s, changes have come at a rapid pace. We can still get by, in many instances, with glues developed 50 and more years ago, on back to possibly 500 years ago. Adhesives such as mortars precede the birth of Christ.

Matching a modern adhesive to a job means almost anything can be bonded and bonded well. Multipurpose glues and adhesives abound today and single-purpose adhesives are generally available for a gratifying number of jobs. No single adhesive works well under all circumstances because the circumstances vary so much.

Conditions and the materials to be bonded affect what material is to be used to create the bond. Some adhesives work well at almost any temperature and some work best at 65 degrees and up. Many adhesives work well only within a narrow temperature range.

Some adhesives must be mixed before use; resorcinol is a fine example of woodworking adhesive that requires mixing. Others come ready-mixed. Others come as a powder (resorcinol comes as a powder and a catalyst that's liquid) such as cascamite resin glues. Some require clamping (most wood adhesives do) while others grab almost instantly. Never as instantly as you would like, I'm afraid. One minute is an awful long time when you're trying to hold things in their proper alignment.

Most adhesives require a clean, dry and well-fitted joint. Others are made to grab just about anything, anywhere (these latter can never be as strong if the joint is not clean).

There are two prime rules for the use of any adhesive or glue. First, always use the correct adhesive for the job being carried out. Second, follow the *manufacturer's* directions for glue or adhesive use. You can almost bond anything to anything these days. Even the little super glues—cyanoacrylates (I'm going to try to keep from spelling that too often because I've got to look it up every time)—have been adapted so they work well with porous materials as well as hard surface materials such as glass and steel. You can now use the glues, or adhesives, for leather, wood, and other such items—if the package is specifically so marked!

Select your glues and adhesives to suit the job. Compare costs and efficiency carefully when a lot of material is required. See Figs. 4-1 through 4-19.

WORKING WITH WOOD

Extensive gluing is required in many woodworking projects. In recent years, a lot of new materials have hit the market. As a group, the wood glues are simple to use, at least in application to the

surface, and in preparation. Conditions of use are another thing.

Ready-to-mix glues require the least fiddling around to prepare the glue itself. Water-mixed glues require only a little more preparation time. Two-part glues and adhesives require that a catalyst be mixed with the base adhesive so that chemical hardening can take place.

Ready-to-use wood glues are the most popular both commercially and for do-it-yourselfers. Why fool with procedures you don't need? Such glues are used for permanent fastening of joints, edge, face, and laminate gluing. There are three types at the top of the popularity list, according to Franklin Chemical's Robert S. Miller. These are the liquid hide glue, aliphatic, and the polyvinyl acetate glues. All set in the same manner with the loss, or evaporation, of water from the glue line.

Any gluing operation consists of applying the liquid glue, dispersed in enough water to make spreading easy to the parts, and then pressing the parts together until the glue sets. As the glue sets, the glue and water will penetrate a short way into the walls of the wood cells. The glue gains strength as the water evaporates.

Fig. 4-2. The Success line of adhesives. Courtesy of Bostik.

Fig. 4-3. Stanley's spring clamps with aliphatic glues.

Speed of water loss affects the speed of set, allowable assembly time, likelihood of wood failure, and depth of glue penetration. The overall amount of wood failure, strength, resistance to water (and other solvents), and heat are dependent on the type of polymer (or adhesive) used.

Viscosity is the measure of resistance to stirring or flow of a liquid. As with motor oil, the higher the viscosity the thicker the liquid, and the more resistant it will be to flow or spreading. Hide glues tend to be almost twice as thick as polyvinyl acetates.

Wet tack is a property that may interfere with brushability while not interfering with other forms of spreading the adhesive. Aliphatic glues have a built-in tack. A benefit is gained when you're working with tight joints because the tacky adhesive is not so easily scraped off as the parts are joined. Wet tack also imparts a bit of strength to the glue before it sets so that light pieces, such as corner glue blocks, will hold themselves in place.

Hide glues are more easily affected by temperature than are polyvinyl acetates and aliphatic glues. If hide glue is used close to its gel point (where the glue begins to become solid) of 70 degrees, it is exceptionally hard to work. Liquid hide glues also lose strength quickly when temperatures approach the 140 degree mark (although they must sometimes be heated so you can spread them).

Aliphatic glues and some of the polyvinyl acetate glues thin out quickly as the temperature rises. The procedure reverses around 120 degrees, and they become so thick as to be unworkable. This

Fig. 4-5. Success cyanoacrylate adhesive. Courtesy of Bostik.

Table 4-1. Glue Comparison Chart.

	Aliphatic Resin	Polyvinyl Acetate	Hide
Looks	creamy	white, clear	clear amber
Viscosity (poises at 83 degrees F.)	30-35	30-35	45-55
pH*	4.5 to 5.0	4.5 to 5.0	7.0
Set speed	very fast	fast	slow
Stress resistance	3 types exceed government requirements of 2.800 pounds per square inch		
Moisture resistance	fair	fair	poor
Heat resistance	good	poor	excellent
Solvent resistance**	good	poor	good
Gap filling	fair	fair	fair
Wet tack	high	none	high
Working temp.	45-110 Deg. F.	60 to 90	70 to 90 F.
Film clarity	translucent, not clear	very clear	clar amber
Film flexibility	moderate	flexible	brittle
Sandability	good	will soften	excellent
Shelf life	excellent	excellent	good

* Adhesives rating lower than 6.0 are acidic.
** Solvent resistance covers alcohols. varnishes. lacquers. and stains taking over the glue joint.
Chart information courtesy of Franklin Chemical Industries

thickness stays with them and the glue can no longer be used.

There are three major types of ready-to-use glues, and a comparison chart (Table 4-1) shows what features each offers.

Liquid hide glues are a modern ready-mix version of one of the oldest types of wood glue. It's still made from the same stuff—hides, bones, tendons—and is far more than adequate for interior furniture uses, forming a joint stronger than the wood itself, with a shear strength on hard maple above 3000 pounds per square inch. Liquid hide glues also work well with other porous materials such as cloth, paper, cloth-backed phony leather, leather and so on, and is frequently used to join veneers.

The glue can be applied using a roller, spreader, pressurized cans and plastic applicators, brushes, or sticks. The rapid set is combined with a long assembly time (whether your project is assembled or disassembled). Most give you two to three hours before setting and some allow as much as eight hours. Set speed can be shortened by increasing the temperature, dropping the humidity, increasing air circulation, or decreasing the spread. Avoid the last choice most of the time. Make sure there's always a light bead of glue that squeezes out when the clamps are applied.

Clamping is essential, but the glue doesn't creep under load and you get a decent-looking light brown to amber glue line. Water and moisture resistance is poor and temperature control (especially on the low side of 75 degrees) is crucial. One of the nicer features is the fact that hide glues do not load up sander belts.

Polyvinyl acetate glues may well be the most commonly used

Fig. 4-6. *Success super glue for porous materials. Courtesy of Bostik.*

Fig. 4-7. *Right; Bostik's Success adhesive line, with the epoxy at the upper left.*

adhesives today (white glues of most types are PVAs). PVA glues serve well in virtually all interior woodworking needs, and are useful for joining paper, cloth, leather, and most other porous materials where a waterproof glue line is not needed. These non-toxic and odorless glues spread smoothly with little or no running or evaporation ahead of time.

Only moderate clamping pressure is required, but it should be

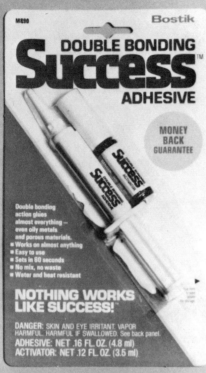

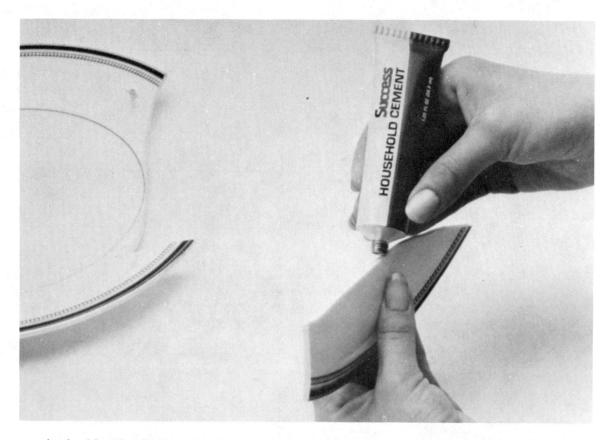

Fig. 4-8. Success household cement. Courtesy of Bostik.

maintained for 16 to 24 hours (a safe bet with any glue or adhesive requiring clamping). Initial set is in one hour for most PVA glues. As with hide glues, clean all squeezed-out glue from glue lines with a damp cloth right away. Later removal is possible but far more difficult.

The glue line is translucent and it will stand moderate stress. Avoid getting the glue on naked steel because it will cause corrosion. Water resistance is a bit higher than hide glues, but not great. PVAs soften and load sander belts.

Aliphatic resin adhesives are still only partially as familiar to most of us as are white glues, but they are among the fastest growing general woodworking glues. Their characteristics are nearly identical to those of white glues, but they are quite a lot stronger and have a faster initial set—20 to 30 minutes.

Clamps are required for only half an hour or so (I usually go for one to three hours, depending on humidity, temperature, etc.), and the glues or adhesives can be used in a far broader temperature range, down as low as 45 degrees, and up as high as 110 degrees. The low temperatures require longer curing time and thus longer clamping times.

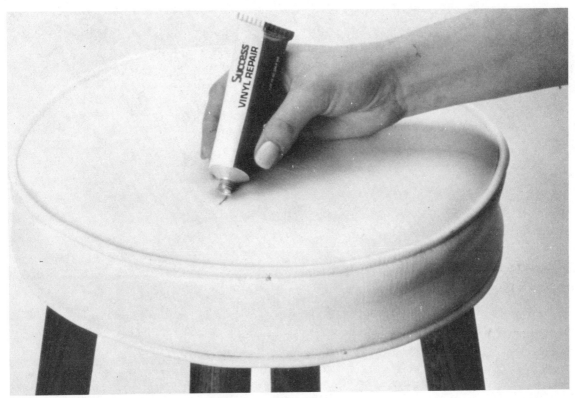

Fig. 4-9. *Success vinyl repair. Courtesy of Bostik.*

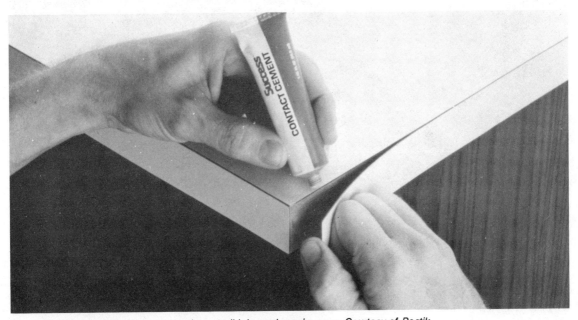

Fig. 4-10. *Success contact cement makes small jobs and repairs easy. Courtesy of Bostik.*

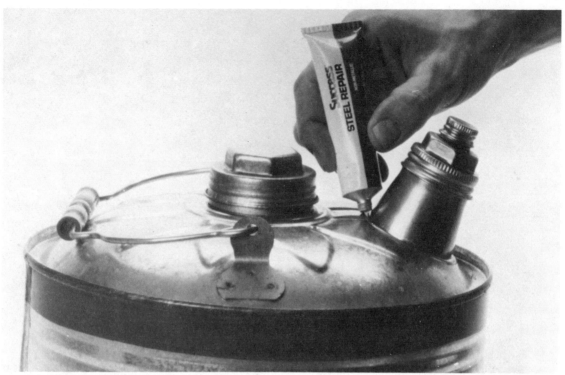

Fig. 4-11. Success steel repair. Courtesy of Bostik.

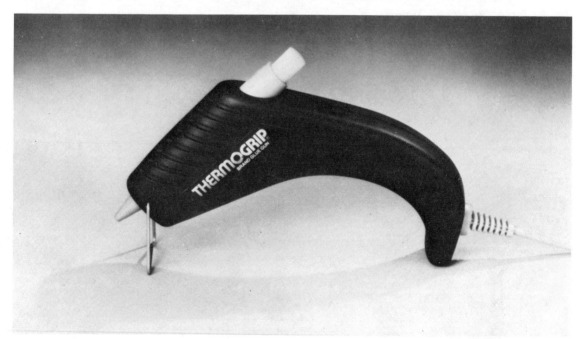

Fig. 4-12. Bostik's solid state Thermogrip hot melt glue gun, #206.

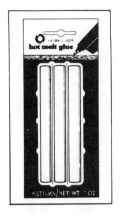

Fig. 4-13. Thermogrip glue sticks.
Courtesy of Bostik.

Heat resistance is good with aliphatics and they have good resistance to solvents. They are easily sanded (something that's not true of white glues).

Aliphatic glues can be colored before application so that color with the final finish is even better. Use water soluble dyes for this. As with the hide and white glues, aliphatic resin adhesives work well on most porous materials, including cloth, leather and paper.

You'll find some aliphatic resin adhesives with what are known as thixotropic features. All this means is that when glue is squeezed out during clamping, it beads at the glue line (or very close to it) without running down the assembled work and making messes in hard-to-get-at corners. As always, clean up with a damp rag as soon as you've got the pieces clamped.

WATER-MIX GLUES

Water-mix glues are those adhesives that come in dry powder form to be mixed with water before use. Mixing proportions of powder to water should be followed carefully, as the manufacturer prescribes.

Water-mix adhesives must be kept in tightly closed containers—I like to transfer mine from the metal cans, with

Fig. 4-14. Thermogrip glue sticks
(boxed). Courtesy of Bostik.

Fig. 4-15. Wide and narrow nozzles for the #207 Thermogrip glue gun. Courtesy of Bostik.

press-on lids (I always seem to bend these things so they don't close as well as they might)—to large plastic bottles such as those used for 1000 pill supplies of various vitamins and minerals. Lumpy and crusty powders, the result of a poorly sealed container, do not hold as well when mixed.

Casein is the old reliable wood glue. It's made from milk protein and comes as a light beige powder. Casein is not waterproof, but is moisture resistant and can be used on exterior work that will be protected so that soaking is not constant. It is nontoxic and handy for making toys because of that.

Casein glues work well with oily woods that other glues will not hold under any circumstances. Teak, lemonwood, and yew are the major woods you might be working with in this area. Casein glue can be worked at any temperature above freezing and it needs about three hours of clamping under moderate pressure for top results. Disadvantages include the fact it sometimes stains some dark, acid woods and its affect on cutting tools (it dulls them quickly).

Plastic resin (urea-formaldehyde) glues, such as Elmer's Cascamite, are light-colored and water-resistant with great strength. They tend to be brittle if gaps must be filled (make sure your joints fit snugly). They must be used at temperatures above 70 degrees and firm clamping for a dozen hours is required for a good bond. The mixture is spread thinly—as are almost all wood glues—and the

Fig. 4-16. The #207 Thermogrip glue gun. Courtesy of Bostik.

joints are clamped. Borden's recommends six hours for hardwoods and five hours for softwoods, with clamp time reduced by half at 90 degrees. The liquid has a mixed life of six hours at 70 degrees. I'd still stay with 12 hour clamping.

Plastic resin glues offer another great advantage. They are just about impervious to molds, mildew, and rot, and leave almost no glue line. You can mix the material with wheat or rye flour (go with wheat flour because rye is more expensive) to make a low-cost, indoor-use veneer glue. For Elmer's Cascamite, use 10 parts flour to 20 parts glue powder to 14 parts water.

Casein glues are creamy in appearance and have a viscosity of 35,000 to 45,000 cps. This can be compared to the tan appearance and 25,000 to 35,000 cps of plastic resin glues. Both are slow setting and both have strength above that required for tests (above 2800 psi) with good stress resistance, moisture resistance, heat resistance, and good solvent resistance.

The casein is a bit better, ranging fair to good, at gap filling, and has a wider working temperature range (32 to 110 degrees as compared to 70 to 100 degrees). Neither offers film clarity, because both are opaque, but the casein is flexible while the plastic resin is brittle in poorly fitted joints. Sandability is good for both and shelf life is about one year.

Two-part glues or adhesives require that you mix a liquid catalyst with the glue resin. In most cases, the working time after mixing is fairly short. Mix only what you need. The price will

Fig. 4-17. Thermogrip products. Courtesy of Bostik.

135

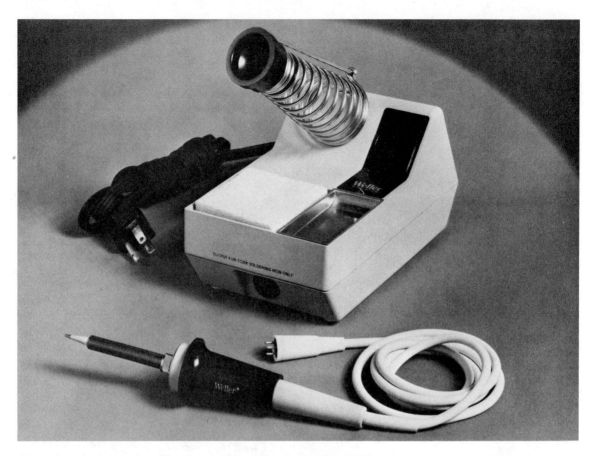

discourage waste anyway. Have everything set up beforehand. Mix according to package directions.

Fig. 4-18. A small soldering iron station. Courtesy of Weller/Cooper Group.

You'll find two types available. One is more widely useful than the first. Resorcinol is an effective bonding agent only on wood. Acrylic resin adhesives work well with wood, glass, and concrete. Neither works on plastic.

Resorcinol Adhesives offer some great advantages for woodworking projects where waterproof glues are required. The strength of the bond is superb while resistance to water is total. Heat and solvent resistance are both good, and the working temperature range is a decent 70 to 120 degrees. Shelf life is about one year. Joints must be tight fitting to prevent brittleness of the film. Sandability is good and set is classed as medium.

The glue line is an ugly reddish brown. Make sure the item is to be painted or the glue stays hidden. The adhesive is mixed using a liquid, cherry-colored resin liquid and a tan powder catalyst or hardener. Measuring cups and other materials for the two parts must always be kept separate. You do *not* want either part of the stuff in your eyes. Fluff the *closed* powder can by shaking and make

measurements accurate. Firm clamping is needed until the adhesive is dry. It will take about a dozen hours at 70 degrees and half that or less at 90 degrees. Leave the clamps in place for two dozen hours anyway. The mix cleans off easily with warm water when liquid, but is virtually impossible to remove once dry.

Acrylic resin adhesives use both liquid and powders, as do resorcinol types, providing an extremely good bond that is also waterproof and unaffected by gasoline or oil. Drying and set time is controlled by the amount of catalyst use with a specified amount of resin. Three parts powder to one part liquid might set in five minutes at 70 degrees. Acetone is required as a cleaning solvent and you get a tan glue line.

When working with wood, you get better control of set time if you first brush the joints with pure liquid before applying the mix. Acrylic resin adhesives are too fast setting for work in larger areas, but they are great for repairs and small work where great strength must be combined with waterproof qualities.

Neoprene construction adhesives (panel and siding adhesives) are carpenter's tools. and as such are only vaguely involved with jointmaking as a process in woodworking. They are exceptionally helpful in constructing stronger homes and other structures of great size.

Fig. 4-19. Soldering gun. Courtesy of Weller/Cooper Group.

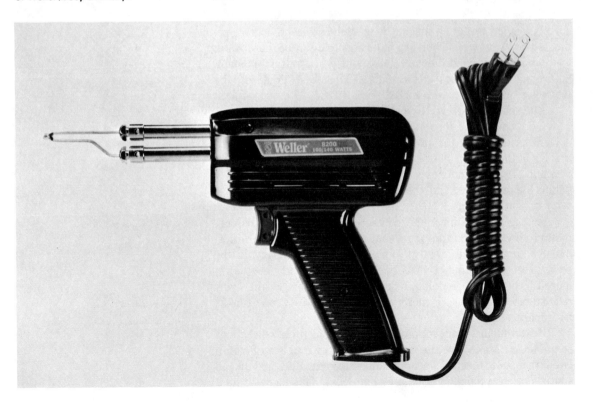

CONTACT ADHESIVES

Contact adhesives are cements applied to two surfaces to be bonded together. The adhesives are then allowed to air dry until you note almost no tack or stickiness, at that time they are brought together. The contact occurs and you had better have everything lined up because the pieces are now joined! Period.

Franklin Chemical Industries states that such cements generally account for some 28 percent of all consumer adhesive sales. They are about the single most popular type for do-it-yourselfers. The main reason for the popularity is the wide variety of uses. In addition to the most common use of bonding laminates to countertops, contact adhesives are used to bond hardwoods, particleboards, leather, cloth, cardboard, plastic foam, ceramic parts, steel, aluminum, plastic and ceramic tile cove bases, gypsum wallboard (though there are easier methods), weatherstrip, auto trim, and so on.

Contact cements come in three basic styles or types: solvent, chlorinated, and water-base. In most cases, solvent-based contact adhesives are best avoided in the home shop. Their very fast dry time—five to ten minutes—manages to keep them on my shelves for warm weather use. I'll only use them outdoors or, when a project can't be moved indoors with all windows open.

Green or start strength is extremely high. No prolonged pressure is needed to get an instant bond. Solvent-based contact cements work very well in bonding laminates to plywood, hardboard, and particleboard, and they are greatly heat and water resistant. Spray and brush grades are available. The brush grade can be trowelled or rolled on with no ball up.

The solvents used are exceptionally volatile and can be toxic.

Only one of the three available Elmer's contact cements is solvent-based. Even the professional cabinetmaker's style is an acrylic latex, though it does require a special solvent for best clean-up. The solvent used with Elmer's is toulene, a remarkably flammable substance, and the cans for cement and thinner also carry poison warnings.

Among other things, the contact cement contains toulene, methyl ethyl ketone, and petroleum distillates. I doubt the formulae differ all that greatly from company to company in the solvent-based cements. Unless one of the special features is required for a job you're doing, forget it. Go with one of the many nonflammable versions. Elmer's makes two and UGL and Franklin both make versions.

Be careful in use and follow *all* of the manufacturer's directions to the letter. Make sure the area, no matter where, is well ventilated, that static electrical sparks cannot be generated, and that all sources of open flame ignition are shut down. This includes pilot lights on stoves, gas water heaters and other such devices. Do *not* smoke anywhere in the same building. All electric motors should be shut down—including those on freezers, refrigerators, etc. All of these precautions must be implemented until you're certain all accumulated vapors are gone.

These things are not a joke. One of my earliest memories involves my father, as a volunteer fireman, coming home sick and disgusted because a foolish young man a few streets up had been working with such materials near a coal furnace during the winter. He never fully recovered from his burns.

Chlorinated-base contact adhesives are less likely to go up in your face. They provide very quick drying, excellent green strength, and ease of application (brush, trowel, roller, or spraying). They smell like a swimming pool in the hot sun, but are nonflammable and are gaining great popularity. These contact cements dry to the touch in 10 or 15 minutes or even 5 minutes under some conditions. The chemicals used include alkaline hydroxide (try lye). Keeping them out of your eyes and off your skin as much as possible is a good idea.

Water-base contact adhesives, sometimes also called neoprene latex base, are nontoxic as well as nonflammable. They work exceptionally well with plastic laminates and gypsum wallboards. They are also recommended for use with polystyrene and other plastics that might be attacked by the chemicals in other contact cements. Use on metallic surfaces is never a good idea. Resistance to heat and water is great. They can be applied with a brush, a stipple roller, a fine-toothed trowel, or a spray gun, they will cover in one coat unless the material is very, very porous (canvas is a good example).

Check the manufacturer's directions and test for yourself on a spot that won't show. The water-base materials won't bother already finished surfaces, and spills can be sponged off while the material is still wet. If the cement hasn't dried, you can use water to clean your tools. If it has dried, take the tools outdoors and clean them as you would any other contact cement applicators. Use a toulene based cleaner/thinner.

Water-based contact cements have one major disadvantage (that will sometimes become an advantage). It takes them about an hour to air dry before contact can be made. Water-base cements are not freeze and thaw stable. Temperatures below about 40 degrees break the emulsion. Contact with even weak acids can precipitate the resin and ruin the bond. Pressure with a roller will be needed and the needs will be for longer and higher pressures than any other contact cements require.

Working with *contact adhesives* is a relatively simple procedure in many ways. Generally the pieces are fairly open and easy to get to

139

(which certainly is not always the case in working with adhesives in general). In most cases, the work done will be lamination, either of wood veneers (check the Woodworker's Store, 21801 Industrial Blvd. Rogers, MN 55374 for their version of heat-applied adhesive) or of plastic laminates meant for countertops and so on. Most often a plywood, particleboard, or waferboard core will be used. Plywood is used when strength is required and the other materials are used for those times when strength needs aren't so great and you want to hold costs down.

The laminate is cut to fit the plywood or other material. You'll find that carbide-tipped cutting tools are best here. The laminates are so tough that they rapidly dull standard steel-tool edges. Use a router, trimming template, and laminate bit for the best results on edge joints and where you must have a very smooth, tight joint.

Apply the contact cement according to the manufacturer's directions, allowing both coats—one on the underside of the laminate and one on the top side of the surface it's to join—to air dry before pressing them together. Make sure the temperature is at least 65 degrees and the humidity under 80 percent, but over 35 percent, before application of the cement. Edge strips usually go on first. The top strip then covers the joint and gives a smooth appearance to the overall joint.

There are several ways to align the laminate before the cemented surfaces touch and bond. The bond is immediate so alignment must be perfect before touching. As a check, you can use a piece of clean kraft paper to see that the surfaces are not still damp. When it doesn't stick even lightly and doesn't pick up any cement, then things are ready to go.

By the same token, I prefer to overlay several sheets of clean Kraft paper cut to 1 inch larger than the surface, and extending out on only one side, on the base. The laminate is then set on the paper with one strip being gently removed as you press gently on the laminate to keep it from shifting. Once this first strip, the smallest of the strips, is removed and that section is bonded, you just pull the rest of the strips loose and press down hard.

If needed, a roller—known as a J-roller—is used to increase the strength of the immediate bond. press as hard as you can with the roller. Try to hit all spots of the surface. You can't press too hard (unless the cabinet under the top being laminated collapses).

The momentary pressure required to bond contact cements is one thing. Lightness is another; press as hard as you can. With no roller on hand, I have used a section of 2 by 4 tapped with a 16-ounce hammer to get the pressure. The roller is better but the 2 × 4 and hammer work.

CONSTRUCTION ADHESIVES

Construction adhesives (Figs. 4-20 through 4-26) are used to install

Fig. 4-7. Right: Bostik's Success adhesive line, with the epoxy at the upper left.

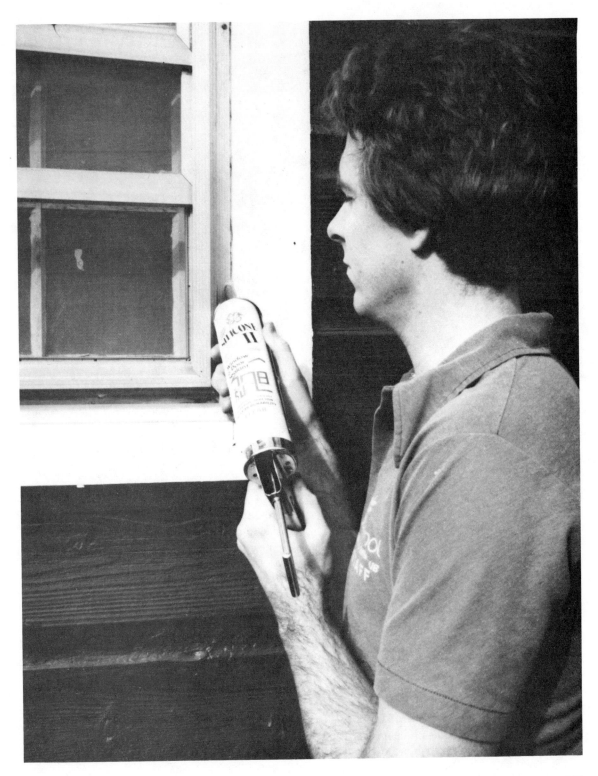

subflooring, wallboard and other items over other subflooring, wall paneling or open studs, depending on whether you've started wth old or new construction. There are a number of types available. Most come in 10- or 11-ounce cartridges for use with a standard caulking gun (some also come in larger, 29-ounce cartridges for use in commercial-sized caulking guns). There is a great versatility to this type of adhesive. Its development led to the American Plywood Association developing its nail/glue system for floors and other areas.

You can install acoustical tile, some bathroom fixtures, brick, and brick veneer (I would still recommend that appropriate mortar in many instances), carpet (indoor and outdoor), ceramic tile (again, the appropriate mortar is usually best), cork, cove base (for plastic tile), hardboard, a wide variety of panels and wallboard, and rigid foam insulation products. Keep in mind that some plastic foams deteriorate quickly with certain chemicals on their surfaces. Adhesives generally meant for such uses are best.

Construction adhesives can be applied under a wide range of conditions, down to about 10 degrees, and even in slight dampness. Drying is very fast at normal temperatures (60 degrees and above finds a drying time of 10 to 30 minutes, depending on the type of adhesive used).

Initial tack is high so a panel is less likely to smack you in the back of the head as you press it up and turn and reach for the next one. A few nails are still recommended, especially with both very

Fig. 4-21. Construction adhesive being used for the American Plywood Association's glue /nail system. Courtesy of the American Plywood Association.

light panels and very heavy panels. The light panels are less likely to twist and warp. The heavy ones are less likely to pull off.

Application depends on the brand and the type, but generally a ¼-inch bead is considered fairly standard on studs (use ⅛-inch beads on sound, solid surfaces), and a 29-ounce cartridge will give 89 feet there. A 10-ounce (or 11-ounce, depending on who makes it) tube will give 32 to 34 feet. You'd get, for a ⅛-inch bead, 355 feet from the big cartridge, and 130 or so feet from the smaller tube.

These adhesives can be used in rough areas where things may not be as fine as you would prefer and with more surface grit and such than is recommended. I can give a qualified yes on that.

When working with construction adhesives, remove dust and other foreign matter from the surfaces to be joined. Apply the correct bead to surfaces. Use beads down studs and a snakelike figure on solid surfaces. Use the adhesive only on one surface. Press the surfaces together and use a few mechanical fasteners to keep them pressed together until bonding is final.

Panel adhesives are used to bond drywall, light wall panels, and so on to studs, other drywall, concrete or, other masonry. You eliminate nail popping problems and patching needs and cut out any chance of messing up prefinished panels with hammer marks. You also help to overcome structural deficiencies because the panel ties the wall together better with adhesives than it can with nails. The application procedures require clean surfaces, reasonable temperatures, use of mechanical fasteners (that can later be removed), and the bonding within 20 or 30 minutes.

Fig. 4-22. Plastic lead fills seams quite well. Courtesy of Genova, Inc.

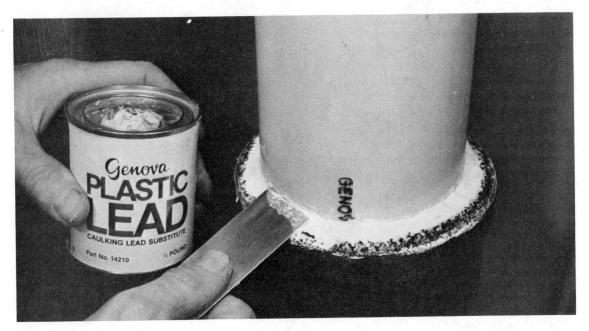

Fig. 4-23. Not a repair product, Genova's Solvent welding cement joins polyvinyl chlorides and chlorinated polyvinyl chlorides by fusing the materials. Courtesy of Genova, Inc.

If the material can be applied with a notched trowel or with a cartridge gun, it's usually some form of construction adhesive. All you really need is a check on whatever water resistant qualities might be needed for a particular job. After that the section is simple and up to you.

Use some care in selecting an adhesive for use with rigid foam insulation boards. These foams will generally be either polystyrene or polyurethane. The polystyrene is somewhat more vulnerable to attack from chemicals in the adhesives. Polyurethane foams, unfortunately, are considerably more expensive than the polystyrenes and are therefore used less often even through the R value per inch is a fair amount greater.

To make sure which kind you're using, if the material is not otherwise marked, figure that a urethane will generally have a yellowish color, with a fine pore structure, while the styrenes are usually white or blue, with the pore structure varying all over the lot depending on who makes the things.

If no other recourse exists, make sure you use a water-base adhesive or test a small corner of the material to see what happens. Give the test spot a full day of exposure.

SPECIALTY ADHESIVES

No adhesive does everything, no matter what the TV pitchmen say. Follow the instructions provided with adhesives. Start with a clean, dry surface but note that some types of adhesives—specifically

Fig. 4-24. Right: Just make sure joints match well before assembling with solvent cement. Courtesy of Genova, Inc.

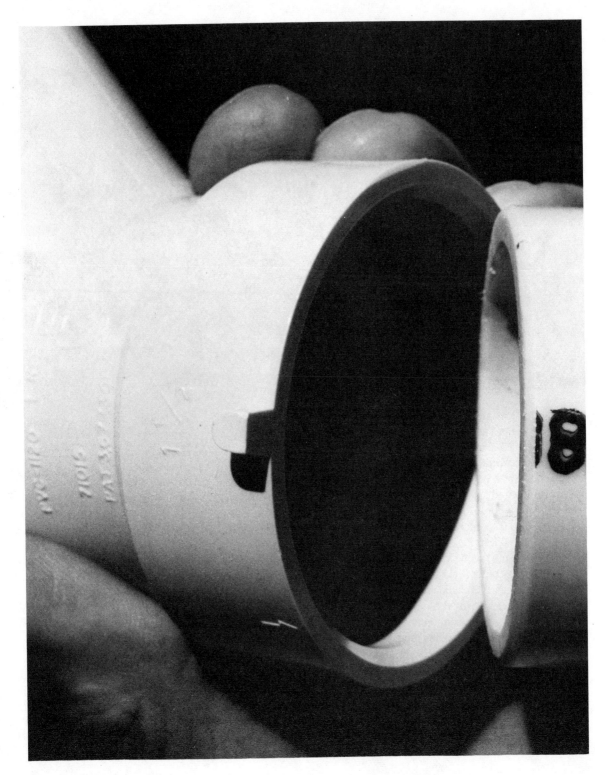

145

mortar for bricks and block—work better when surfaces are damp. Make sure that all grease is gone no matter what kind of adhesive is to be used.

Cyanoacrylates are called miracle adhesives and super glues, and in many ways they fit the bill. In other ways, the advertising that has accompanied them since their introduction has justifiably soured many people on their use. Setting time is exceptionally quick at 10 to 30 seconds. This makes them very susceptible to drying in the container.

When the quick set time is added to the fact the cyanoacrylates are anaerobic—they do not need air to cure—the in-the-tube drying phenomenon was increased. Containers have recently been redesigned so this is less of a problem. The glues won't dry readily in the presence of oxygen and the containers now reflect this fact.

Until recently, such glues did not work at all well with porous materials. Elmer's has recently come out with a leather and wood version. On metal, glass, rubber and most plastics, the grip is almost instantaneous and powerful. Most of my success with them has come on such surfaces.

Temperature resistance is good, up to about 160 degrees, and most chemicals have about as much affect as water—which is none. Acetone cleans them up, fortunately, for the glue will bond skin

Fig. 4-25. Check alignment before cementing. Align just a fraction of an inch off as you assemble, twisting to final marks immediately (you've only got a few seconds of working time, I promise). Courtesy of Genova, Inc.

together in a rush. Cyanoacrylates are expensive so use over large surfaces is not practical. Always have some acetone (nail polish remover) on hand when working with these things. Even that doesn't work after about three minutes of set time! Adhesive sets in 10, 15, or 30 seconds, depending on the maker, and a full cure is obtained in six hours.

Epoxy adhesives were first developed for industrial needs, including construction of aircraft components, and are now generally used to bond many hard-to-glue materials such as glass, metal, porcelain, concrete, and a number of plastics. Because hardening is by catalytic action rather than by evaporation of the solvent. Epoxies will harden inside an airtight container or, for that matter, under water.

Epoxy resins provide some of the qualitites of adhesion and bonding needed, but overall bonding is more heavily affected by the type of catalyst used to get the cure or set. Most resins used by consumers are of the polyamide resin group because they're easier to use than many other kinds. The proportions mixed are not as crucial as with other catalysts. A good cure can be guessed at even when you "eyeball" the mix. Adding hardener will give the cured bond more flexibility. Catalyst resins usually work best at a 50/50 mix, but they can be easily varied to 60/40 either way with no appreciable losses. As you increase the epoxy resin in the mix, the

Fig. 4-26. Apply the cement generously but not sloppily. Courtesy of Genova, Inc.

mix tends to become brittle from the lack of catalyst. Nevertheless, bond will still be good.

Epoxies in clear and colors are widely available. All provide strength well up into the thousands of pounds per square inch category for shear (depending on hardener type and proportions in the mix). Resistance to solvents is exceptional and most epoxies serve well as water-resistant fasteners even though they're not meant for use as waterproof glues (another common mistake in use). They hold well under a wider range of temperatures than do most other consumer adhesives, but failure is common at uses that require continued exposure to more than 200 degrees Fahrenheit. Continuous temperature cycling from hot to cold has a bad effect on the bond.

Cost for wood joints is generally out of line. The material does not have anywhere near the flexibility needed for many types of wood joints. They are useful for repairs to leaky pipes and for ceramic tile-fixture installation. They're much stronger than ceramic tile mortars when you're going to install soap dishes, towel bars and such, but many of the current crop of ceramic tile mortars are now epoxy reinforced. Other uses include fixing wrought iron railings without welding, repairing metal, plastic and furniture, and so on.

Epoxy formulas come in two packages. Tools and storage containers must be kept separate just as with resorcinol mixes. With typical epoxies, no clamping is required after a thin coat is applied to the surfaces to be joined and they are pressed together. No dimensional changes occur as they dry. You don't need to worry about shrinkage. The strength of bonding is very high. Nevertheless, you might want to tie or tape the parts in place until the epoxy cures. Generally, you can expect epoxies to set in four to six hours. Using a heat lamp—controlled for a temperature of no more than 250 degrees—will cut cure time to one hour or less.

Epoxies in bar form are readily available these days and are very handy for many things. Two bars are wrapped separately: one resin and one catalyst. You cut equal amounts—with different knives—off the bars and knead them together until the color is uniform. After that they are applied to the parts to be joined or sealed.

This stuff is great for sealing small holes in various kinds of containers made of metal where welding or brazing or even soldering could be hazardous. I've used it on gas tanks for a riding mower, on a tank for a tiller and on two large, expensive gas cans (beats buying replacements, and saves the purging time welding would require).

Simply wipe the hole area with alcohol—to remove any grit, grease and oil—and apply the epoxy putty. If you prefer, you can lightly sand the area to ensure a better bond. This is a particularly

strong, and will also take the extreme hot water beating a dishwasher hands out. In addition, the newer cements are nonflammable and the fumes are not harmful.

Surfaces must be clean and dry and the cement is then applied to a single surface only. Position the pieces closely without touching them together. Once positioned, press them together for at least 60 seconds. After that, check for pressed-out glue, and wipe away any excess with a clean cloth. They dry to a clear finish in two to five (bet on five) minutes.

If you have more pieces to add, give the first piece or two time to cure. Then let the assembled part stand overnight before running it through your dishwasher. These cements will stand dishwasher heat but they will not take oven heat.

Acrylonitrile adhesives are also called buna-N adhesives and were developed for use on aircraft. Buna-N adhesives are a one-part, thick, tannish liquid that give a flexible, totally waterproof bond useful for fastening fabric to metal and glass and wood. They work well in bonding patches to repair clothing and rips in carpets, tents and other items. The adhesive gives greater strength than sewn repairs.

Of the several ways of applying buna-N adhesives, wet bonding requires you to apply the material to both surfaces to be joined and you must wait until it dries to a tacky condition. Then you push the parts together with enough pressure to get a good contact until it cures.

In some fabric repairs, the material is applied to the patch and allowed to dry completely. Bonding takes place when the fabrics to be repaired are heated and pressed with a flat iron on low heat (about 200 to 300 degrees, depending on the exact formula used).

Acrylonitriles can be applied with what's known as a reactivation method. As in heat and pressure bonding, the adhesive is applied to one surface and allowed to dry completely. When a repair is needed, the buna-N is reactivated by wetting it with a solvent prior to application. Solvents used are acetone, ethyl acetate, and methyl ethyl ketone. Hand pressure is then used for joining the parts. In most cases, the reactivation method is not used in do-it-yourself applications.

Buna-N adhesives are not suitable for wood-to-wood bonding. Nevertheless they work well in bonding other materials to wood.

Fabric mending cements are meant to eliminate or reduce the need for sewn repairs in wool, cotton, canvas, leather, denim, and many other fabrics. They are used both with and without a supporting patch to make repairs in many items. When correctly used and dried, the patch will withstand normal washing, ironing, and other wear procedures.

Fabrics must be clean and dry for proper repairs. A small

good idea with some paints, but use care to use an abrasive material that doesn't throw sparks. Don't sand so hard you heat things the stuff surely does eat through the synthetics).

Polyester adhesives sound as if they should be used in the garment district, but they are used to bond many materials, with the widest use in construction and repair of fiberglass boats. The material is a two-part adhesive, with a liquid resin and a liquid catalyst, mixed (as are all two-part adhesives) just before use.

Usually, you'll require just a few drops of catalyst to a good-sized dollop of resin. Too much catalyst will harden the mix before you can apply it. Because the various formulae change hardening times drastically, there's no way to give accurate times here. Suffice to say you should follow the manufacturer's directions closely. Remember that heat applied shortens curing time.

Polyvinyl chloride (PVC) adhesives are useful for bonding glass, china, metal, plastics, and other such items (hard plastics only). It can also be used on wood and other porous materials and is a one-part adhesive. Apply it evenly over a single surface (making sure the surface is clean and dry) with a two-or three-minute wait following to allow the PVC adhesive to become tacky.

Press the parts together until things are set, unless the parts are liable to move out of place. Use clamps or some other method of fixing them if that's the case. If the surfaces are porous, apply the PVC adhesive to both surfaces and let it dry to touch before joining the parts. Cure takes place in 20 to 30 minutes. Clean-up is with acetone or lacquer thinner.

Cellulose cements, cellulose nitrate cements, or nitrocellulose cements are usually called household cements or model glues. These dry clear and take a set in 2 to 10 minutes, with only a little pressure required on the joint. A complete cure takes 24 hours. These cements are flammable when wet.

Cellulose cements will join wood, metal, leather, paper, glass, and ceramics, but will not withstand a great deal of stress and are only moderately water resistant. There's some shrinkage as they dry and that can help to draw parts together.

A thin coat is applied to each surface and allowed to get slightly tacky before the parts are joined. Hand pressure for two or three minutes will then give a good bond in porous materials, but nonporous materials take five minutes or so. Acetone will remove any excess, but don't spill the cement on any painted surface or fabrics. I don't remember it harming natural fibers such as cotton or wool, but the stuff surely does eat through the synthetics).

China and glass cements are a development to replace the cellulose cements in one specific job (as aliphatics and others have taken over other jobs). There are a number of different formulae for china/glass cements these days that set clear and hard, are very

amount of the cement is then spread on both surfaces and the pieces joined with light pressure. Wait 24 hours after application before using or laundering. For laundering, iron the repair for one minute at 300 degrees (wash-and-wear on most irons) after the cement has dried two hours. Of course, allow the repair to cool before use. Water cleans up the cement before it cures. Alcohol will remove the excess later.

As with all other such cements and adhesives, test for fabric compatibility before using the material. Some fabric mending cements are latex based (drying by evaporation). The resulting bond is flexible and rubberlike and will withstand repeated laundering.

Plastic mending cements are in ever-increasing demand these days because of our increased use of plastics in general. Specific cements serve specific purposes for repairing flexible plastics and rigid plastics, whether of styrene, vinyl, acrylic, or phenolic resin bases. The clear, flexible adhesives can be used to repair plastic raincoats or other sheet materials, plastic toys, and plastic swimming pool liners. They are also good for assembling plastic models and repairing beach toys and other such gear. Some can be used on wet surfaces and even under water. Pool liners can be patched without a need for draining.

Most flexible items should have a patch made of a material similar or identical to the material being repaired. The patch should be large enough to extend at least ½ inch beyond the edges being repaired. Lightly sand or otherwise roughen the surface of both the patch and the area being mended to get a better tooth for the adhesive. Apply the cement all around the edges of the patch and over the center as well if there is material left in the to-be-repaired area the patch can contact. Press the patch in position and clamp with light-to-moderate pressure for about three hours or until the cement dries. Next, let the patch dry overnight before use.

Plastic rubber cements are available from many sources in a wide range of formulations. Basically, the stuff is a sealant with a strong adhesive. It is used mainly to repair rubber items permanently. It usually has a creamy or paste consistency, is made of neoprene rubber, and is found in a translucent amber as well as black and white. It mends rubber items and can be used with insulated tool handles. Plastic rubber cement will adhere to almost any clean, dry surface, is not much dependent on porosity—so it is useful for bonding all sorts of dissimilar materials—and it is useful for sealing as a caulk in wood, metal, and around windows and so forth in homes.

Plastic metal cements, as with many adhesives, come in a variety of formulations. Most, such as plastic aluminum, contain a vinyl resin with atomized metal in either liquid or putty form for ready use as a filler or adhesive. They take a metallic luster when polished or sanded after they're cured. They can be worked in much the same way as the metal they repair (from drilling through filing and sawing).

The surface to be repaired needs to be clean, free of dirt, oil and grease, and should be roughened with abrasive cloth or paper for best adhesion. Apply the cement and level it. After three or four hours, it will have had time to set and can be worked. If the application is thick or deep, apply the material in ⅛-inch layers for best results.

Rubber-based adhesives such as styrene-butadiene will be one-part, water-resistant glues for bonding to nonporous materials such as metal and glass. It is useful in replacing loose tiles and bricks, attaching bathroom and kitchen wall fixtures, and holding nuts and bolts together without lock washers. Gap filling properties are excellent, and the adhesive can be drilled, tapped, sanded, and painted when hard. This adhesive can even serve as a sealer when applied under water for emergency repairs. Surfaces need to be clean and tight before application. You then butter the adhesive on one surface, allowing it to air dry until it becomes tacky. No clamping is normally needed once the two surfaces are joined; mineral spirits or turpentine will clean up any mess.

Silicone adhesives were once unusual, and remain moderately costly, but serve well as sealants and caulking compounds. They considerably extend the life of residential and other caulking jobs over most other types (some are now advertised as good for 30 years, although 20 is still more common). The material is packed in squeeze tubes or in cartridges. It dries to a flexible consistency that neither cracks nor dries out. Chemical resistance is excellent and resistance to heat and cold is good, whether the material is used indoors or outdoors.

Look for the silicone sealer for a specific job. The various formulations are often made with enough difference so that performance characteristics are far better in one situation than another—intentionally. Setting time is not rapid, as it might be with other types of adhesives, but the material works well on glass, ceramics, metal, china, and a large number of plastics. A few types are made especially for use on wood, canvas, and ruber.

Concrete bonders are acrylic resins designed for application in one of five ways. Follow directions to the letter.

As an adhesive tie coat—a concrete bonder—allows you to bond new stucco, plaster, or cement mix to old concrete, stone, wood, or metal. The bonder is spread over the surface to be repaired or recoated using a spray gun, brush, or roller. It is not to be diluted unless the label specifically tells you to do so; I don't think many do. You've then got about two hours to get the new concrete or whatever down and over the bonder.

As a fortifier, the bonder gives added strength to Portland cement and dry ready-mix mortars and cements. It is added to the dry concrete mix before water is added. You must follow instructions on the package closely for decent results.

Concrete bonder can also be used as underlayment to level off uneven floors of old concrete, wood, old tile, and such materials. It can be troweled (mixed as if for a fortifier) from a featheredge to up to 2 inches thick, as the leveling requires. It will fill all cracks and voids and give you a smooth surface to work from.

As a primer-sealer, the bonder is applied to porous materials such as gypsum wallboard, plaster, concrete, and stucco before you paint or grout the materials. When used in this manner, you'll usually mix the bonder two parts to one part of water and apply it evenly, using a brush, roller, or spray gun. Give it about 45 minutes to dry and you can then apply a water-thinnable mix. You've got to wait until it's completely dry before paint application.

As a decorative nonslip coating, the bonder is used in many places such as around swimming pools, on all sorts of walks and steps, boat decks, docks, and on metal, wood, or old concrete. Use a concrete ready-mix and bonder mixture in just about the same proportions as described for fortifier use. Spread a 1/16-inch to ⅛-inch-thick coat, and then use a stiff bristle brush or broom to make a rough texture in the surface. You can add acrylic paint colors to the wet mix if you prefer.

Hot melt glues are either thermoplastic polyamides or polyethylene adhesive based, and are used in an electric glue gun. The adhesives come in stick form about ½ inch in diameter and 2 to 4 inches long. My gun is by Bostik and is called Thermogrip. The fancier model has three nozzle shapes and a trigger.

These glues are good for bonding many porous materials, are nontoxic, nonflammable, odorless, and pretty much not bothered by moisture. About the only real disadvantages are the fast set times once the glue is melted.

The use is simple, with gun loading being very straightforward and easy, with about a three minute gun warm-up time. Keep your fingers off the nozzle and heat chamber because they get quite hot—right up on 400 degrees.

Remember that once glue flow starts you've got about 10 seconds of working time. Don't work large areas at any one time. The bonding surfaces must be clean—have you read that before?—and the adhesive will gain 90 percent of its holding ability within 60 seconds of heat removal. These guns are excellent for any number of jobs, and the newer guns also feature the ability to use a hot-melt caulk, with a 60-second set time. The caulk also makes a good, waterproof adhesive and sealer!

Paper adhesives start with mucilage and going through rubber cements, library pastes, and various animal and vegetable glues. Most are applied with spreaders or brushes. The two surfaces are then pressed together.

My most recent adventures in such work are with the Craft Bond materials from Elmer's. Craft Bond I and Craft Bond II are inexpensive to use for decorative and other kinds of paper work. In most cases, white glue and wood glues work as well. They are a bit more costly and will require more care in use.

If you want a permanent bond, use one of the white glues. If you want a bond that will release with little or no destruction of the paper, use a rubber cement. As always, clean and dry surfaces work best, and the manufacturer's directions are the ones to follow.

Chapter 5

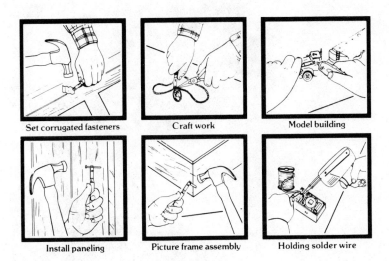

Set corrugated fasteners | Craft work | Model building

Install paneling | Picture frame assembly | Holding solder wire

Mortar

Brick is usually considered an expensive building material, but the highest portion of the cost results from what one must pay a masonry expert to lay the material. Even for an expert bricklayer the material is labor intensive. The same holds true for many forms of ceramic tile. See Figs. 5-1 through 5-9.

There's a cure. Lay your own brick, block, or ceramic tile. Much of the secret is in the measuring and cutting, and the rest lies in the mixing of the various mortars required for the jobs in different areas. You'll certainly never get as fast as the person who does the work every workday, but you'll be fast enough to save much money and add to the value and attractiveness of whatever you might be building.

If you think you might enjoy the work, give a small project a try. Remember that brick is heavy to work with and the mortar is cumbersome to mix.

Bricklaying has its own jargon. Brick faces have different names and the methods of laying them up are easily discernible. Buying materials, as well as working with them, is easier once you know the vocabulary used.

There are several patterns in which brick, block, or tile can be laid. The adhesive mortar bond is only a small part of what keeps a brick wall together, in most cases, so check bond styles for those most likely to hold or help hold things together. See Figs. 5-10

through 5-29. The stack bond should always be avoided in any wall that must support or help support more than its own weight. Other bonds are generally suitable for any use. Header courses might be required if double wythe and thicker walls are used.

A *course* is a continuing row of bricks. When bonded to other courses, they form the masonry structure.

A *wythe*, or continuous vertical, 4-inch (or wider) section or thickness of masonry, is probably best explained as a vertical row or course.

A *stretcher* is a brick laid flat with the longest part parallel to the face of the structure being built.

A *header* is a masonry unit (brick or block) laid flat with its longest side perpendicular to the face of the unit being erected. Headers are most often used to tie two wythes together in solid brick masonry. They are also used for decorative purposes in brick veneer.

A *rowlock* is a brick laid on its face.

A *bull-stretcher* is a brick laid with its longest face parallel to the wall.

Fig. 5-2. Glazed quarry tile from Summitville's Lombardic line is used for this enclosed patio. Courtesy of the Tile Council of America.

Fig. 5-3. Left: American-Olean tile is
used in a kitchen and dining area.
Courtesy of the Tile Council of
America.

A *bull-header* is laid as a rowlock with its longest side perpendicular to the wall.

A *soldier* is a brick laid on its end so the the longest dimension is parallel to the vertical mortar joints of the wall.

JOINTS

After the mixing of the mortar, the mortar joint is the most important part of a good, solid brick or block wall. The joint starts with correct hand position on the trowel. The thumb rests at the base of the handle (not circling the grip). Generally, the trowel will be pointed down and away from your body, with right-handed workers holding the trowel so the left edge is used to pick up mortar. Left-handed workers reverse that. Start by picking up just enough mortar to cover three bricks for the first dozen or so tries. You'll get too little or too much but your eye will soon get the hang of things.

One-brick coverage will form a thin row along the trowel edge. Enough for five bricks will fully load the standard trowel. Hold the edge of the trowel directly over the center of the brick course and tilt the mortar into place. Move the trowel to your right (again,

Fig. 5-4. Astec's Texlace tile used
for a sunken tub. Courtesy of the Tile
Council of America.

Fig. 5-5. Ceramic tile makes an excellent surround for planters. Courtesy of the Tile Council of America.

Fig. 5-6. Florida Tile's Potter's Touch makes a superb kitchen countertop. Courtesy of the Tile Council of America.

Fig. 5-7. Uneven or rubble quarry tile from Universal Ceramics fits well with the brick wall for this patio. Courtesy of the Tile Council of America.

left-handed workers must reverse the procedure). Scrape off any mortar hanging over the course edges and return it to the mortar board.

Spread the bed of mortar until it's about 1 inch thick, with a furrow up its center, tapering from the center of the brick. Butter the end of the first brick and shove it into the mortar bed. When a brick is picked up to be buttered and bedded, your thumb will be on one side of the brick and your fingers on the other. Keep one end in the air and butter that end. Buttering is just what it sounds like.

Place as much mortar on one end of the brick as it will hold. Too little mortar placed during buttering means your head, or vertical, joint will be open and not resistant to water penetration. Push the brick into the bed and up against the preceding brick, if there was one, hard enough to force mortar out of both the head and bed joints. Cut off the excess mortar as it seeps out.

Use a mason's line and a straightedge to keep the joints horizontally aligned. Check corners frequently for vertical alignment. At the outset and until you gain quite a bit of experience, level every two or three bricks.

Fig. 5-8. Brick line 4- ×-8 inch tile from Summitville gives a country look to a dining area. Courtesy of the Tile Council of America.

Fig. 5-9. Wenczel's Village Square tile sets off this bathing platform quite nicely. Courtesy of the Tile Council of America.

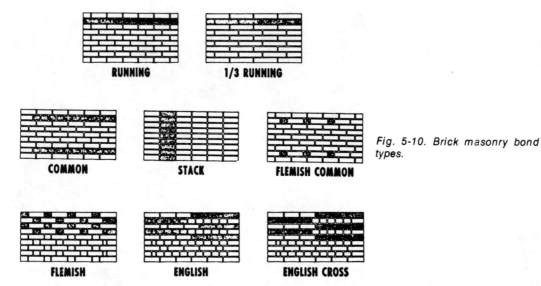

RUNNING   1/3 RUNNING

COMMON   STACK   FLEMISH COMMON

FLEMISH   ENGLISH   ENGLISH CROSS

Fig. 5-10. Brick masonry bond types.

This works best for two reasons. You can do the leveling before the bricks are starting to set into the hardening mortar and you can also catch any leveling errors before they have a chance to build up and throw the entire structure off. Leave the "eyeball" leveling to those with much experience.

After the mortar in the joints is set up thumbprint hard—to the point where it will just take a thumbprint under moderately hard pressure—you need a jointing tool and you come down the joints with the tool to produce a concave joint that is solidly packed and dense enough, as well as shaped correctly, to help keep moisture out and preserve its integrity.

That is the basic chore of laying brick or, on a moderately larger scale, concrete block. Blocks are heavier and a little harder to handle individually, but they go up far more quickly as far as area covered is concerned. As far as numbers are concerned, you should be able to lay several bricks in the time it takes to insert a couple of 50-pound-or-so blocks.

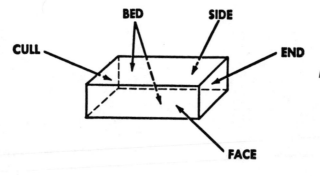

BED   SIDE

CULL   END

FACE

Fig. 5-11. Brick surface names.

164

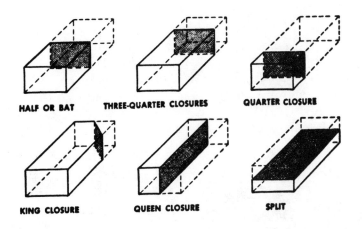

HALF OR BAT   THREE-QUARTER CLOSURES   QUARTER CLOSURE

KING CLOSURE   QUEEN CLOSURE   SPLIT

Fig. 5-12. Cut brick shapes.

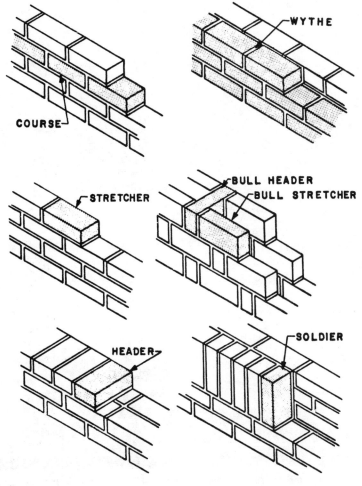

WYTHE

COURSE

STRETCHER

BULL HEADER
BULL STRETCHER

HEADER

SOLDIER

Fig. 5-13. Masonry unit names.

165

Fig. 5-14. Proper way to hold a trowel.

Fig. 5-17. Mortar is thrown on brick.

Fig. 5-15. Picking up mortar.

Fig. 5-18. Spread over three to five bricks.

Fig. 5-16. Positioning mortar before throwing it on the bricks.

Fig. 5-19. Cut off excess mortar.

STEP 1

Fig. 5-21. The bed joint furrow should be close to the center of the mortar.

Fig. 5-20. Make a shallow furrow in the mortar bed.

Fig. 5-22. Buttering a brick.

MIXES

The correct mixing of mortar—in dry mode and wet mode—is all important to success when constructing anything of brick or block. If a mortar is mixed properly dry, just enough water is added, and the joint is then pointed properly after the joints are well filled as the bricks are laid, then the wall should last, and last, and last.

I doubt anyone knows just how long modern mortar mixes, based on portland cement, will really last, but I would think a well-done job should let you stare at things from under an angel's wings, no matter how young you now are. Tuckpointing repairs should last just about as well.

Concrete in its usable form has plasticity and is therefore readily molded. It changes shape only slowly when a mold is removed. The speed of the shape change is governed by the mix of the concrete. This is also true with mortar.

Concrete is concrete, cement is cement, and mortar is mortar; except it isn't quite that way until the entire job is finished. Portland cement is the base for all three items. Uniformity of mix is of great importance and the type of mix is also of importance. Mortars are more or less resistant to various forces depending on what has been used to make them (and in what amounts).

Bond strength of the mortared wall is affected by the type and

Fig. 5-23. Poorly bonded brick.

quality of the cement used, the surface texture of the bedding areas (how rough or smooth the brick surfaces are), the plasticity of the mortar, and the water retention of the mortar, and the quality of the sand used for the mortar.

It really all sounds far more complex than it is. Simply put, use top-quality portland cement, top-quality lime, good, clean sand, and clean water. Use no more water than is required after getting a thorough dry mix of the proper ingredients for the conditions your wall is expected to withstand.

Mortar in its correct wet form is just plastic enough to be worked with a trowel without segregating. Segregating can be caused by too little water in the mix, too much sand, or dirty sand. *Sharp sand* or *mason's sand* is a good bet, but it is not as widely available as we might hope these days and it is more expensive than regular sand.

Stop adding water well before you've completed the mixing and check for plasticity as you mix. It's far easier to add water than to try

Fig. 5-24. Setting the brick.

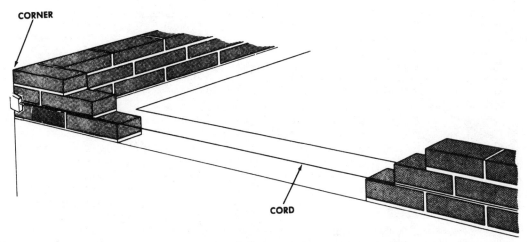

CORNER

CORD

Fig. 5-25. Using mason's cord to maintain alignment.

to mix in more ingredients if the mix is too slushy. If the mix tends to run out of the joints and down the brick faces, you'll have weak mortar joints and an ugly looking job. That stuff is terrible to get off once it cures.

Masonry cement can be used to produce mortar, but to be honest I haven't seen any in years. Portland cement is readily available just about everywhere and masonry cement is nothing but portland cement with lime added. Portland cement does just fine.

For standard, moderate, or light-stress use above ground, you need an ordinary service mortar with 1 part portland cement and ½ to 1¼ parts of hydrated lime, plus 6 parts of sand in damp, loose condition. Masonry cement will require no lime and only 2½ to 3 measures of sand (all measures of materials are by volume here).

For severe-service mortar joints, in those areas with violent winds or severe frost, use 1 part of masonry cement, 1 part of portland cement, and 4½ to 6 parts of sand. Or use 1 part portland cement to a quarter-measure of hydrated lime, and only 3 parts of sand.

Fig. 5-26. Mortar joint styles. Use only concave styles for exterior masonry.

For designated mortar types M, S, N, and O, you can really spread the uses. Type M mortar is for use below ground and for

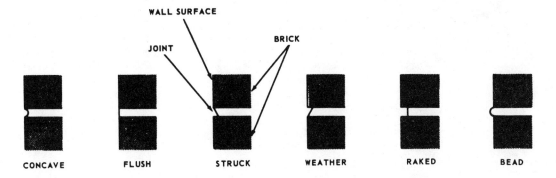

WALL SURFACE

JOINT

BRICK

CONCAVE   FLUSH   STRUCK   WEATHER   RAKED   BEAD

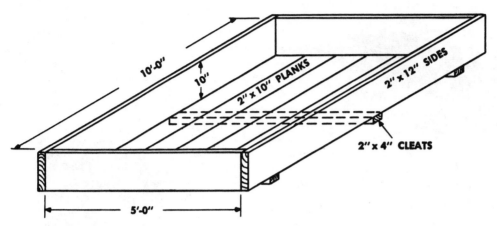

work in direct contact with the ground. It is especially useful in retaining walls. Use 1 part portland cement, ¼ part hydrated lime, and 3 parts of sand.

Type S mortar is a general service type. It provides very high resistance to lateral forces—high winds and so on. Use 1 part portland cement, ½ a part of hydrated lime, and 4½ parts of sand.

Type N mortar is used when weather conditions are really rotten and where wind and blowing water might contain high concentrations of salt, such as along our coasts. Make it with 1 part portland cement, 1 part hydrated lime, and 6 parts of sand.

Type O mortar in most cases is classified as light duty and should be used only where load bearing walls do not carry more than

Fig. 5-27. Mortar box plans.

Fig. 5-28. Mortar mixing board plans.

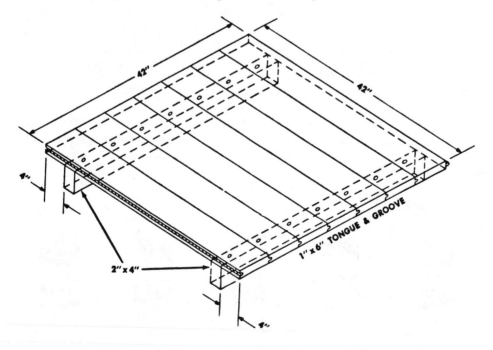

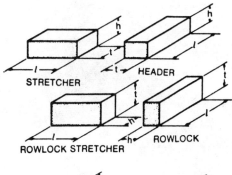

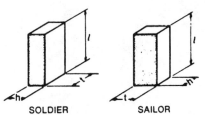

Fig. 5-29. Brick positions.

STRETCHER

HEADER

ROWLOCK STRETCHER

ROWLOCK

SOLDIER

SAILOR

10 pounds per square inch of pressure. Type O is not strongly resistant to the freeze-thaw cycle and not as resistant to mortar as most other types. Use 1 part portland cement, 2 parts hydrated lime, and 9 parts sand.

Mortar is mixed in much smaller amounts than concrete. A mortar box is inexpensive to buy or easy to build, or you can use a wheelbarrow. Mix all dry ingredients very carefully before adding any water. I usually use a garden hoe, but special hoes with holes in the blades do a slightly faster job (and are still useful in the garden if cleaned well).

Mortar that has stiffened somewhat can be retempered by adding a small amount of water and giving it a thorough remixing. It is often hard to tell the difference between the initial set and stiffness from water evaporation. Time is your best ally here. If the temperature is 80 degrees or higher and the mortar has been mixed for over two hours, discard it instead of tempering. If the temperature is above 60 and below 80 degrees, you get a 3-hour time lapse.

If you have any doubts at all about the mortar, discard it instead of retempering. The very few dollars you might save if the retempering works is not worth the chance of a failed bond in your brickwork.

CERAMIC TILE

Modern tiles come in ungrouted and grouted sheets as well as individual tile shapes. Very important in any tile job is the mortar selection. The number of jobs that can be done, as shown in Figs. 5-30 through 5-42, is wide. Just a short list includes tiled tabletops,

Fig. 5-30. This bath platform uses tile from Mid-State Tile Co. Courtesy of the Tile Council of America.

Fig. 5-31. This sunken tub also has a greenhouse effect. Courtesy of the Tile Council of America.

Fig. 5-32. Basic tile job needs for a planter. Courtesy of the Tile Council of America.

tiled countertops, tiled floors and walls, and decorative tile used around a fireplace opening.

From there, you can move to Wonder Board tiled and set out from the wall as a protector when a wood stove is used for heat, and you can build and tile a platform for the wood stove to make it a decorative room feature. Planters can be tiled and you can box in not-so-attractive home details (such as garbage cans) and tile the box.

You can tile the floor or cabinet bottom under your kitchen sink. Spills of waxes, detergents, and other stuff are then easily wiped up. Other projects include home bar tops, barbecue tables, walks, stairs, and areas of patios (or entire patios can be built of paver tiles). Swimming pools tend to run costs out of sight, but you might want to tile a fish pond or a yard fountain.

For more information, write the Tile Council of America, DIYS, P.O. Box 222, Princeton, NJ 08540. Enclose half a buck and ask for their two tiling leaflets or a list of their ideas.

With modern adhesives, grouts, and cutting materials, tiling is less of a problem than ever before and quite readily within the realm of do-it-yourselfers. By doing it on your own, you save the largest cost of any ceramic tile job and bring a luxury material into a reasonable cost range. Care is needed to obtain the best-looking job possible, even with modern materials, but the completion will be very satisfactory.

You'll find you have most of the tools required for a job of tiling already on hand if you do any other around-the-home work at all. You need a straightedge, a tape or rule, a torpedo or 2-foot carpenter's level, a square, a carborundum stone (for smoothing cut tile edges), a scraper, a sponge, and some clean rags.

Special tools needed are a tile cutter or a good set of nippers (both of these can be rented or borrowed from the dealer who sells you the tile), a notched trowel, and a rubber trowel or a squeegee.

The process is a four-step one once you have measured things and determined the tile layout. Start by spreading the adhesive with the notched trowel, and then set the tile. After about 24 hours, apply the grout, and then clean the tiles and polish them if needed.

Cutting tiles is the part that makes most people nervous, but with a rented guillotine tile cutter the tile is measured, marked, set in the cutter, scored, and broken off. Tile nippers take out those little bits you need to round corners and let pipes through. With a carborundum stone, smooth the cut edges that are to be visible. That's it. More or less.

Fig. 5-33. Cut the mesh mounts for the mosaics so that they fit the planter, top to bottom. Courtesy of the Tile Council of America.

Fig. 5-34. Spread adhesive with a notched trowel. Do only a small section at a time. Courtesy of the Tile Council of America.

Fig. 5-35. Use a light twisting motion to press the sheet of tiles in place. Courtesy of the Tile Council of America.

Fig. 5-36. Complete setting tile. Courtesy of the Tile Council of America.

Fig. 5-37. Cut out the border design. Courtesy of the Tile Council of America.

Fig. 5-38. Fill in the design with contrasting mosaics. Courtesy of the Tile Council of America.

One of the greater complexities in doing a good tiling job is the selection of both setting and grouting materials. Choose right on those two and all goes well all the way through. Choose wrong and things are unlikely to stay in place.

Grouts are no longer the simple white glop set into the spaces between tiles. Summitville alone offers 10 color choices from snow white to solid black, and including country gray, avocado, natural, caramel, palomino, and a number of browns. This allows for an entirely different look even when the same tile is used in various areas of a house or other structure.

S-100 is a setting and grouting epoxy mix for floors and walls. This is a three-part mix made to resist attack from acids and alkalis and other such compounds that would normally eat great holes in grout and mortar. It's fairly costly, but easy to mix and very sag resistant (important on walls). It can be cleaned with water and it is nonflammable as well as non-toxic. No linear shrinkage is listed, and the stuff has very good shear strength after 14 days: 271 psi. The initial set takes seven hours.

S-200 epoxy adhesive, for setting only, is a two-part epoxy mix for both floors and walls. It has an easy-to-mix feature, high bond strength, and good adhesion on virtually all structurally sound surfaces, including existing ceramic tile. It's a bit lower in cost than some adhesives, and it has a tan color. There is no linear shrinkage and shear strength (bonded to plywood) is 213 psi after seven days. Cure time is about 16 hours, sometimes less at room temperature, and the material has a working time of 3 to 3½ hours, with no sag.

S-300 is another epoxy adhesive for setting only. The two-part product is essentially the same formula as S-200 with more epoxy resin and catalyst to obtain a higher bond strength. The color is tan. The shear strength after seven days on plywood rises to 315 psi. The working time is about four hours.

S-400 is a setting and grouting epoxy, two-part, with a severe service recommendation for use in distilleries, chemical labs, food processing plants, and so on. It's kind of costly for home use, but it is great for setting floor bricks, quarry pavers, and ceramic mosaics. Shrinkage is nil and bond strength is an incredible 1,156 psi. Cure time is 12 to 16 hours, with a one-hour working time, and no sag. No visual deterioration is seen after 28 days of immersion at 72 degrees in a 48 percent solution of hydrofluoric acid.

The same resistance extends to a long, long list of chemicals,

Fig. 5-39. Mix the grout. Courtesy of the Tile Council of America.

Fig. 5-40. Apply the grout with a squeegee. Courtesy of the Tile Council of America.

including a 45 percent solution of sulfuric acid, and a whole list of things.

S-759 is a thin-set mortar for most floor installations. With high bond strength, it has about a 40-minute open working time. Figures for resistance to shear and such are not readily available for such mortars, but his should serve for almost any residential use. It has a companion, in S-763, that is meant for walls. It is sag resistant and has the same open time.

S-1000 is a thin-set latex mortar mix with a 329 psi shear bond strength after 20 freeze and thaw cycles. It has a 28-day dry cure strength of 360 psi with mosaic tiles. Sag is .023 and open time is up to half an hour. Water resistant qualities are really great because the material is made for use even in water-saturated environments. It is used on both floor and wall tile installations.

Grout is not just a joint filler. It aids the overall strength of the installation by keeping water and chemicals from reaching the mortar or other adhesive used to install the tile. Unsanded joint fillers (grouts) are used when joints are narrow or when you use a porous bodied or glazed tile.

Compressive resistance and stain resistance are good and it can be found in white, beige, and dark brown. Even ink leaves only a slight, removable stain. Fungus- and mildew-resistant grouts are unsanded and used in narrow joints with porous bodied tiles. Molds simply cannot grow; this includes black mold. It's available only in white most of the time. Mustard, coffee, and ink do stain but are easily removed.

Sanded joint fillers are used on both floors and walls. They should be made to give a dense, tightly packed joint, with good bond strength. These are the most generally used grouts and come in the widest range of colors (all 10 at Summitville). Often special colors are available on order. Seven-day compressive strength is high, at 5,335 psi, and the initial set takes one to four hours—with a final set in three to four hours. For general tile use, this is your best bet. Unsanded grouts are fine for countertops and mosaics where joints are likely to be narrow. The lack of the bulky sand makes them easier to get into the narrow joints.

Grout clean-up is usually done with a wet sponge, after the grout has been spread over the face of the tiles and forced into the joints. Some pregrouted sheets are liable to require a special grout

Fig. 5-41. Wipe off excess grout. Courtesy of the Tile Council of America.

Fig. 5-42. Plant and admire the results of your project. Courtesy of the Tile Council of America.

of flexible, usually silicone seal base, material used from a cartridge. These might require cleaning with alcohol. If that's ever the case as you're working or if another volatile clean-up material is recommended, make sure there is no open flame and plenty of ventilation.

Tile setting is hardest, I think, in materials selection. Once the choices are made the work is far easier. Wood, clay, and plastic tubs are available most everywhere. Except for the brassbound redwood types, they look awful all the time and even worse once the flowers bloom and provide a contrast with the planter.

It's easy enough to paint some of these planters, but paint tends not to last. Covering a planter with ceramic tile, using mosaic tiles on sheets, simplifies the job. The mesh or paper backing and the small tiles will follow almost any contours with little trimming.

Use a thin-set adhesive for the job. I'd go with one made for water-saturated environments. Rainwater collecting in the dirt won't soak the adhesive loose ruining your job.

You shouldn't need a tile cutter, but you can rent one if necessary. Make your design by lifting the solid color tiles off the backing. Later install the varied colors. This is most easily done after the sheets have been cemented in place. Apply adhesive to the tub using a notched trowel, covering only a small area at a time. Cut the sheets to fit from top to bottom, and place them with a slight twist as they go down.

The border design requires little more than lifting out or cutting out two rows near or at the top of the sheets. Fill those rows with a constrasting color and make sure the joints are aligned well.

Wait 24 hours and mix and apply grout with a squeegee. Use a sponge to clean excess grout off the face of the tiles and a clean rag to polish things up an hour or so later.

Fill and plant and watch things bloom.

Most American-made mosaics come from American Olean, Lone Star Ceramic Company, and Winburn Tile Mfg. Company, but lines change with some frequency. For other decorative uses, you might well find other tiles in various catalogs. Check with your local tile dealer.

Chapter 6

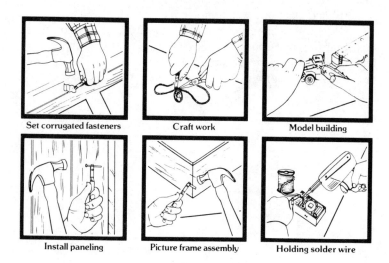

Set corrugated fasteners   Craft work   Model building

Install paneling   Picture frame assembly   Holding solder wire

Metal Fasteners

The fastening of metal seems a maddening job to a great many people, but in reality it's often a lot easier than fastening wood. Much depends on the selection of fastening materials you choose to use. Metals are often subject to a greater variety of stresses than are wood items. Most of us neither need nor want a full-scale metal forming shop. Those of us who do get involved in more extensive metal work still want to keep fastening and forming costs to a minimum.

Nuts and bolts and screws and rivets form the greatest array of metal-fastening items today. These fasteners can be installed, if care is used, by almost anyone with any degree of experience at all (Fig. 6-1). Heat fastening of metal requires at least modest bits of practice as well as fairly expensive tools.

SCREWS AND BOLTS

This section describes removable and reusable fasteners for metal, screws, and bolts. Screws (Fig. 6-2 through 6-7) are probably the most common method of fastening metal parts. *Machine screws* are screws used in tapped holes or with nuts for metal part assembly. The threads of the tapped hole or nut must match the screw threads. With machine screws being made in many diameters, lengths, pitches (number of threads per inch), materials, head styles,

Fig. 6-1. Working with metal fasteners. Courtesy of DRI, Inc.

finishes, and thread fits that can be something of a selection problem. The simple selection of a half-inch machine screw is compli-
cated by need for material specification, thread specification, head
type specification, and so on.

A typical job spec might include ½ inch, 8-32 roundhead, steel,
blued, machine screw. The first number is screw length, but the
8-32 might be a mystery. The 8 is the screw gauge number, or
diameter, and the 32 is the threads per inch.

Machine screw materials and finishes can vary all over the lot
with so many companies custom making screws for special jobs. In
most cases, the screws are likely to be aluminum, steel, or brass.
Steel screws are often plated to help prevent or cut down on
corrosion. Occasionally, even brass machine screws are plated.
You can often find screws made of special materials such as
monel—(highly resistant to corrosive action of saltwater, among
other things). The screws in my left knee are made of monel.

Head shapes for machine screws tend to be even wider in
variety than for wood screws because the applications are likely to
be wider. There are nine common styles in use and a dozen or so in
general use for specific purposes. Some of those you are not likely
to encounter.

Some heads require special driving tools that might or might
not be included with the screws when you have a unit requiring
assembly. In most cases, if the screw head is quite uncommon the

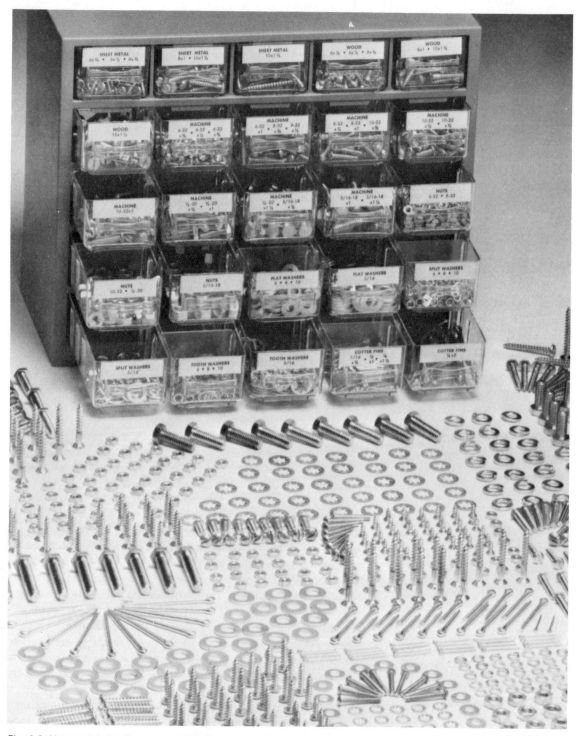

Fig. 6-2. Nuts and bolts. Courtesy of DRI, Inc.

WOOD SCREWS			MACHINE SCREWS						SELF-TAPPING SCREWS		
	DRILL PILOT HOLE				DRILL THROUGH HOLE		DRILL TAPPING HOLE			DRILL PILOT HOLE	
GAUGE DIAMETER	BIT NO.	DIAMETER INCHES	GAUGE DIAMETER	*THREADS PER INCH	BIT NO.	DIAMETER INCHES	BIT NO.	DIAMETER INCHES	GAUGE DIAMETER	BIT NO.	DIAMETER INCHES
0	75	1/64									
1	71	1/32									
2	65	1/32	2	56NC	43	.089	50	.070			
3	58	3/64	3	48NC	37	.104	47	.079			
4	55	3/64	4	40NC	32	.116	44	.086	4	42	3/32
5	53	3/64									
6	52	1/16	6	32NC	27	.144	36	.107	6	39	3/32
7	51	1/16							7	35	7/64
8	48	5/64	8	32NC	18	.170	29	.136	8	32	7/64
9	45	5/64									
10	43	3/32	10 / 10	24NC / 32NF	9 / 9	.196 / .196	25 / 21	.150 / .160	10	30	1/8
11	40	3/32									
12	38	7/64	12 / 12	24NC / 28NF	2 / 2	.221 / .221	16 / 13	.177 / .185	12	24	9/64
14	32	7/64	1/4 / 1/4	20NC / 28NF	F / F	.257 / .257	7 / 3	.201 / .213	14	12	3/16
16	29	9/64									
18	26	9/64									
20	19	11/64	5/16 / 5/16	18NC / 24NF	P / P	.323 / .323	F / I	.257 / .272			
24	15	3/16	3/8 / 3/8	16NC / 24NF	W / W	.386 / .386	5/16 / Q	.313 / .332			
	Drill this size hole to start screw.				Drill this size hole for a snug fit.		Drill this size hole to tap threads.			Drill this size hole to start screw.	

*Note: NC after machine screw threads-per-inch indicates a coarse thread (National Coarse); NF (National Fine) indicates more threads per inch. More threads per inch gives greater holding power . . . use NF threads for screws and bolts that will be under exceptional stress.

Fig. 6-3. Guide to selection and use of fasteners. Courtesy of DRI, Inc.

driving tool will be included. I've yet to see one that will last much longer than the job at hand (and often not that long).

American Standard Machine screws come in roundhead, flathead, fillister-head (a sort of high oval), ovalhead, trusshead (a type of roundhead, with a flat area at the base of the head and around it), binding head (another form of roundhead, with a binding flat at its base), panhead and panhead recessed, (flathead styles without the taper, with the recessed having a slight arch to its head), hexagon head, and 100-degree flathead (the taper on the sides of the head is near 100 degrees instead of the standard flathead 80 degrees).

Slot styles begin with the plain old standard slot and then get fancy. The one-way slot means a screw requires a special tool for removal after a standard-tip screwdriver is used for installation. Phillips come in several styles, but with the same basic cross pattern, while hexagon-head screws might be slotted or unslotted so that a nutdriver is required to insert and remove them. Twin-head wrench types require a square-tipped screwdriver.

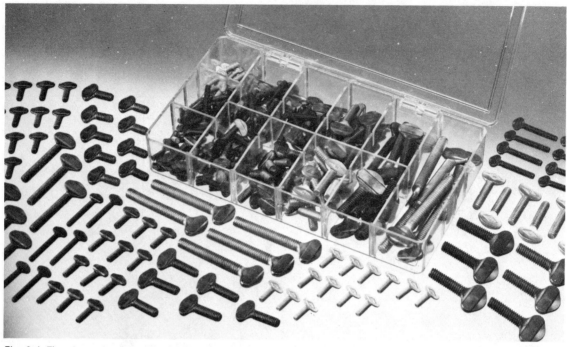

Fig. 6-4. Thumbscrews. Courtesy of DRI, Inc.

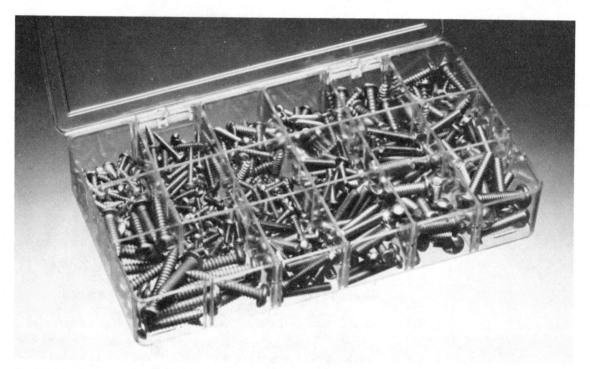

Fig. 6-5. Metal screw variety. Courtesy of DRI, Inc.

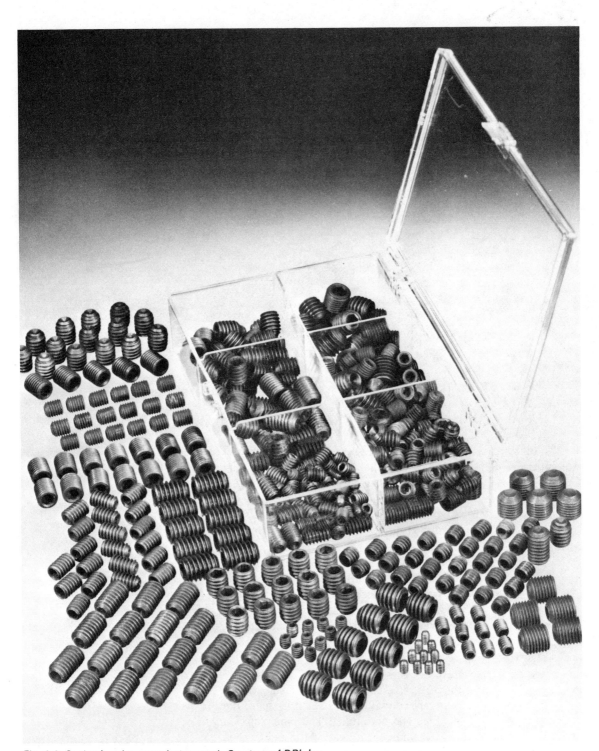

Fig. 6-6. Socket head screws (set screws). Courtesy of DRI, Inc.

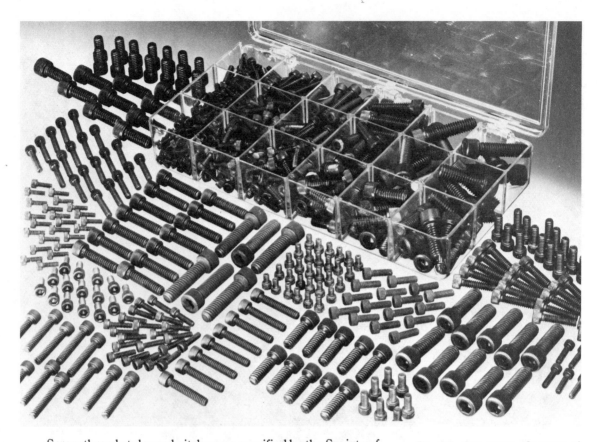

Fig. 6-7. Capscrews. Courtesy of DRI, Inc.

Screw thread styles and pitches are specified by the Society of Automotive Engineers (SAE) and the National Screw Thread Commission has standardized on three thread styles. The SAE Extra Fine is not recognized by the other body, which uses the designations National Coarse and National Fine. The SAE's EF styles are for use in aircraft and autos where such threads may be required.

Fit classes were also designated so that tightness and accuracy are easily determined. Class I is a loose fit; Class II is a free fit; Class III is a medium fit; Class IV is a close fit. Loose fit, or Class I, is probably what you will find in sheet metal shelving kits where a bit of shake is acceptable. Free fit is for parts you might have to put together using just your fingers. A medium fit is for tighter tolerances and would require tools for assembly. The close fit is for fine-grade work, absolutely requires tools and allows little shake or looseness. No wonder the SAE came up with its own designation!

Machine screw threads are cut with tools known as tapes; the common types are taper, plug and bottoming. The taper tap is a starting tap that provides a chamfer length of eight to ten threads that doesn't cut, while plug taps have a clear length of three to five

Fig. 6-8. Right: Castellated nuts. Courtesy of DRI, Inc.

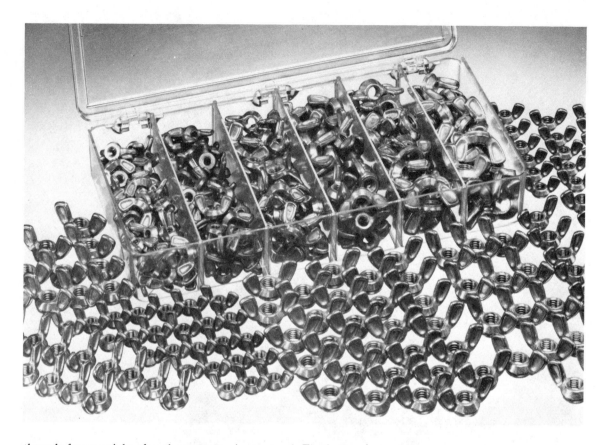

Fig. 6-9. Wing nuts. Courtesy of DRI, Inc.

threads for use right after the taper tap is removed. The bottoming tap has no free length, or only a one thread chamfer at most, and is used to run to the bottom of blind holes. The taps are always used in succession. A mix of lubricant is used for tapping in steel, depending the quality steel. The lube must be thick. Many machinists seem to prefer a half and half mix of white lead and liquid animal fat.

The tap goes in a quarter turn and is then backed out the same amount, with a turn forward to take up the space before making the next cut. This helps clear chips and makes for a cleanly cut thread. You'll need to exert pressure on the tap for the first three threads or so, but after that it will draw itself down and cut with just a twist. If you get a feeling of springiness, back off the tap immediately, withdraw it, and clean the cutting flutes of chips and debris. Blind holes are most likely to cause this problem because there is no place for the chips to go. Back the tap out often and, if possible, turn the stock over and tap it with your hand to remove chips.

Capscrews are similar to machine screws but are available in larger sizes so they'll hold more or tighter. Capscrews are seldom used with nuts. Usually, they'll be cut to pass through a clearance hole in one piece to tap into a second piece and hold the two

Fig. 6-10. Right: Cap (acorn) nuts. Courtesy of DRI, Inc.

192

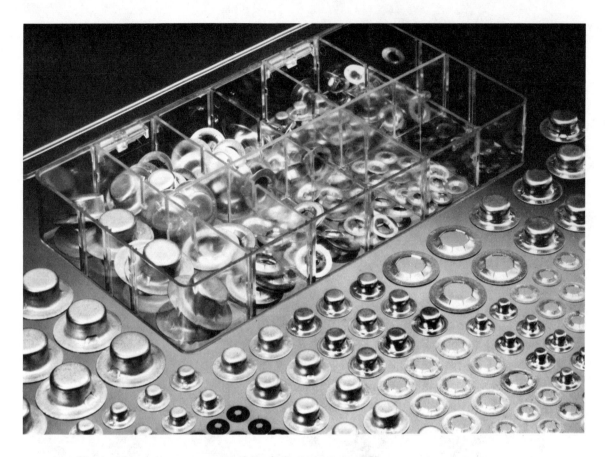

together. The head is an important part of the entire unit; it must be well made or it will pull off.

Threads can be any class of NF or NC, and the heads can be round, square, flat socket or fillister style. The entire screw is well finished in all instances.

Setscrews are those little things that hold doorknobs and other knobs in place on a shaft. They can also be used to hold hubs in place. Setscrews are almost always made of hardened steel. Cones and dog points (more or less square) hold best. The point fits into a matching recess in the shaft or simply bears on the shaft (which tends to wear the shaft badly after a time).

Setscrews are headless or headed in several ways, Allen, hex wrenches, or sometimes a thumbscrew type will be used.

Self-Tapping Metal Screws are a great invention. After you've had to install a few dozen feet of stovepipe for a wood stove, you'll agree—if you're doing a good installation. Each juncture needs at least three such small screws. I used #7, ½-inch screws that are hard to run into the metal with a screwdriver or nutdriver. Chuck the drive in an electric drill and things take far less time and effort.

Fig. 6-11. Push-on nuts. Courtesy of DRI, Inc.

Fig. 6-12. Aircraft-nylon insert-lock nuts. Courtesy of DRI, Inc.

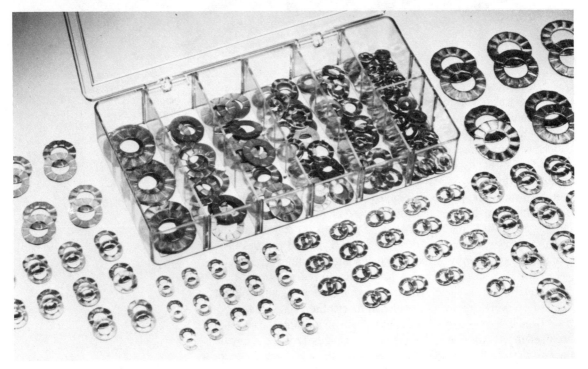

Fig. 6-13. Lock nuts. Courtesy of DRI, Inc.

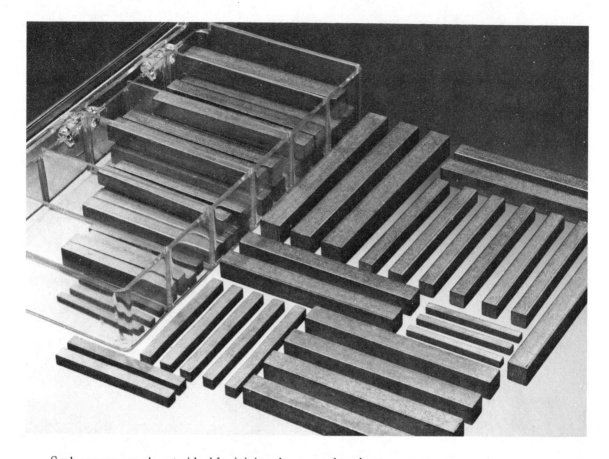

Such screws are close to ideal for joining sheet metal to sheet metal. They also work very well in joining sheet metal to wood and joining some man-made materials such as plastics either to plastic or to metal. As with virtually all screws, there is a good variety to choose from, making many suitable jobs easily done. Type A sheet metal screws have very coarse threads and are used for joining sheet metal from 18 gauge to about 28 gauge, as well as for fastening metal to wood, and plywood to plywood, hardboard to hardboard, etc. Diameters available are about #6 to #14. Commonly available lengths are from ½ inch to 2 inches. If you look hard, you can find them in ¼-inch lengths and down to a #4. Type AB has the very sharp point of the type A but also has somewhat finer threads.

Type B sheet metal screws have blunt points and moderately fine threads. They are useful for thicker-gauge metals, with holes at least started. These are not useful in most other materials.

Type C screws have threads like those on machine screws and are used mostly for joining sheet metal in gauges 12 to 22. They do not cut threads. Instead they force their way through metal, requiring quite a lot of drive torque power.

Fig. 6-14. Keys, square, and rectangular. Courtesy of DRI, Inc.

Type F sheet metal screws do cut threads and they also have fine threads. Use is in aluminum and steel and plastics up to ½ inch thick.

Type U screws are often called drive screws. They are really more metal nails than screws. The shank has a twist like that of a screw nail and they are usually driven with a hammer into solid castings. Sheet metal is attached with them on a more or less permanent application.

Most true drive screws have a hexagon head, but many other types are available. I use a great number of sheet metal screws and try to select, for stovepipe installations, hexagon head types with a small built-in washer under the head. That helps prevent smoke leakage.

In other cases, I select a simple drive screw or a nondrive screw and use it, as I would any other, to fasten metal to metal or to other materials. Their usefulness is often under-rated because people tend to believe they are not suitable for wood uses as well as metal uses or for combined uses. Sheet metal screws of the self-tapping variety are useful in many other materials as well as metals.

Bolts, studs, and *nuts* (Figs. 6-8 through 6-23) might be the most important metal fasteners in heavy jobs for almost all of us. If

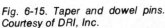

Fig. 6-15. Taper and dowel pins. Courtesy of DRI, Inc.

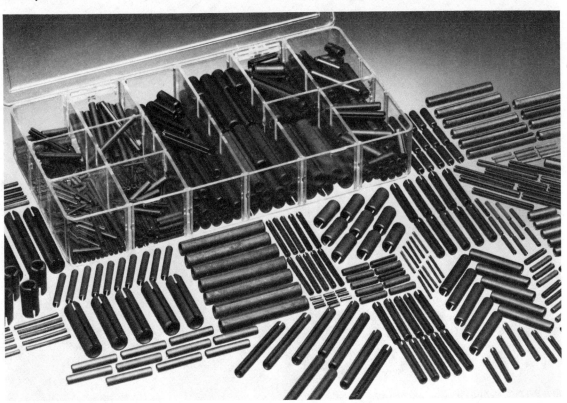

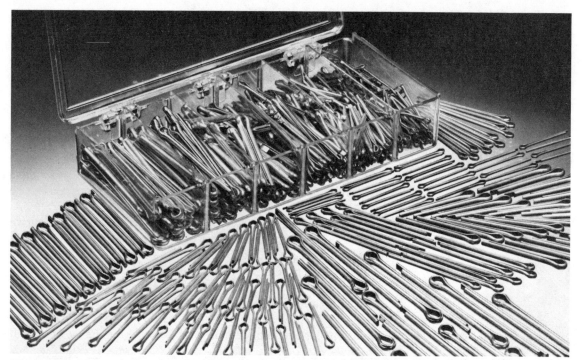

Fig. 6-16. Cotter pins. Courtesy of DRI, Inc.

Fig. 6-17. Tinnerman J nuts. Courtesy of DRI, Inc.

Fig. 6-18. Tinnerman nuts. Courtesy of DRI, Inc.

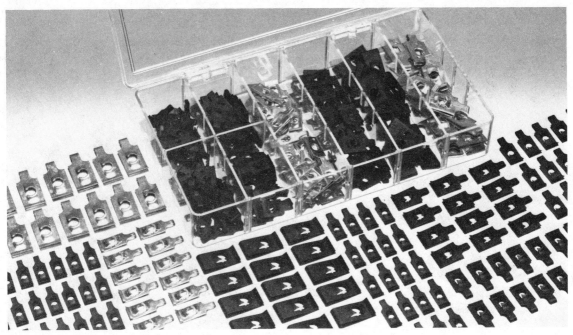

Fig. 6-19. Tinnerman U nuts. Courtesy of DRI, Inc.

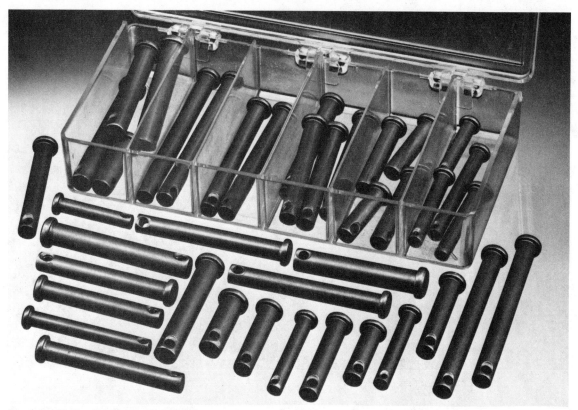

Fig. 6-20. Clevis pins. Courtesy of DRI, Inc.

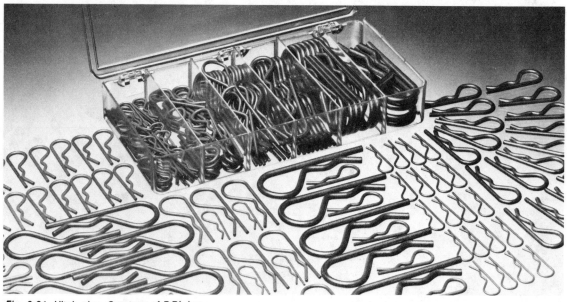

Fig. 6-21. Hitch pins. Courtesy of DRI, Inc.

200

Fig. 6-22. Circlips. Courtesy of DRI, Inc.

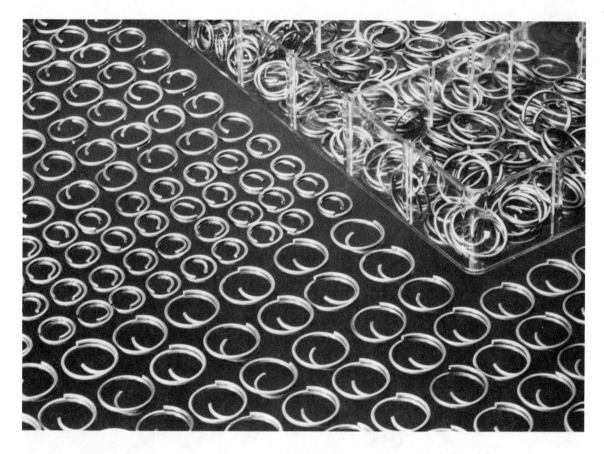

you do any auto work beyond a basic car washing, you will have to learn about metal fasteners.

Roundhead bolts are used as fasteners all the way through items to be held, with a matching nut that may be square or hexagonal. The various types of roundhead bolts include carriage bolts and stove bolts, with several types also useful (as are carriage bolts) in wood construction. Bolts can be hot or cold formed, and they get only the machining needed to form the threads so the overall appearance is often somewhat rough.

Machine bolts have either square or hexagonal heads. They are primarily intended for use as through fasteners with nuts and washers. Machine bolts can also be used as capscrews.

Machine bolts, unlike roundhead types, are available in several finish grades and in a number of strength grades. Use the appropriate grade for your work, but don't go hog wild in adding strength where it isn't needed because the more operations needed to form any bolt the more it must cost.

When you can, install all bolts with the heads up. This allows the bolt to stay in place if the nut and washer should vibrate off.

Fig. 6-23. Ring clips. Courtesy of DRI, Inc.

Fig. 6-24. Right: Stanley's Flip Drive ratchet handle.

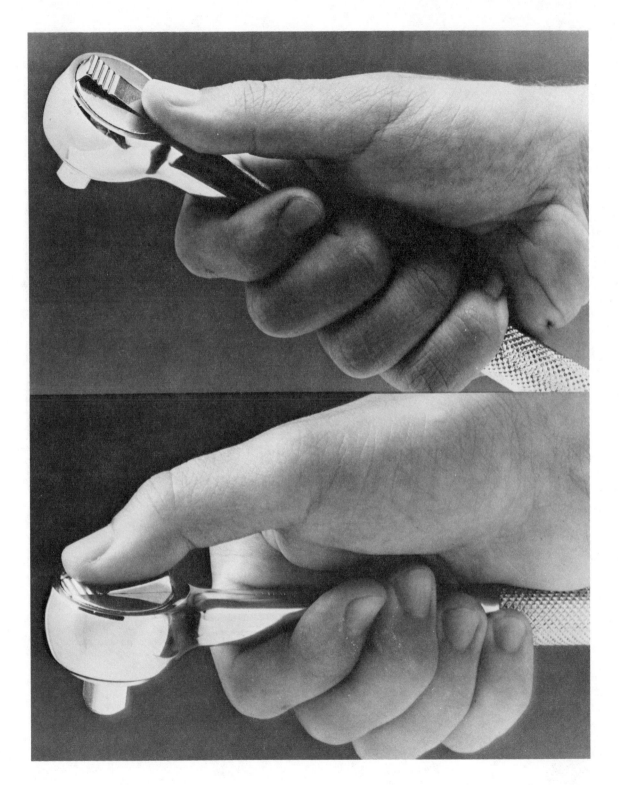

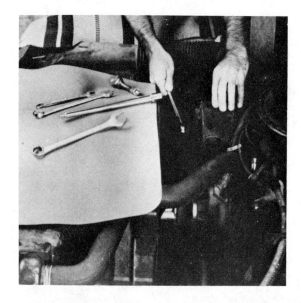

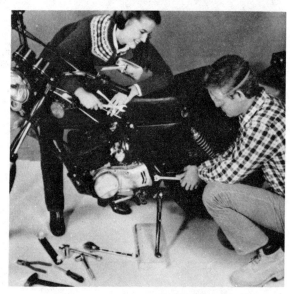

Length, in most cases, should equal the thickness of the materials being joined. The threaded portion of the shank should show about one turn after the nut, with the unthreaded shank just passing through the materials to be joined.

    Studs are rods, usually threaded on both ends. They are used for such things as holding heads on an auto engine. The stud passes into the engine block and then extends through a hole in the head to the point where a washer and nut can be added. The nut draws the parts together for a tight seal. Gaskets are usually used with

Fig. 6-25. Some of the jobs done with wrenches and sockets and metal fasteners. Courtesy of The Stanley Works.

automobiles, but they are not needed if machining of the joined materials is precise enough and the units joined are massive enough to not warp.

Studs can readily be made at home or they can be bought as needed. Custom-made nuts and bolts sounds a bit odd, but you'll frequently find manufacturers with stocks left over, or at least the taps and dies on hand to rapidly cut even small numbers).

Nuts come in as wide, or nearly as wide, a range as do the screws, bolts, and studs they must fit. Castellated nuts have slots up the sides so that safety wire or cotter keys can be inserted through a hole in the bolt end to keep things from loosening up. Castellated nuts will show up in most all vehicles on the front wheels, holding the wheel bearings in place, at the very least.

Jam nuts are used above a standard nut to lock it in place. They differ from standard nuts in that they have a bearing surface at their bases to "jam" against the lower nut.

Wing nuts are used in jobs where only finger tightened pressure is required and where quick removal and installation may be of importance.

Cap nuts are sometimes full nuts, with a nice finish, and sometimes only caps that fit over regular nuts. Either way they are

Fig. 6-26. Vaco nut driver set.

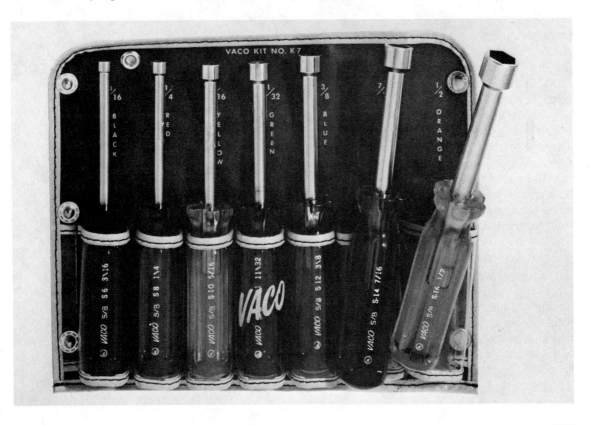

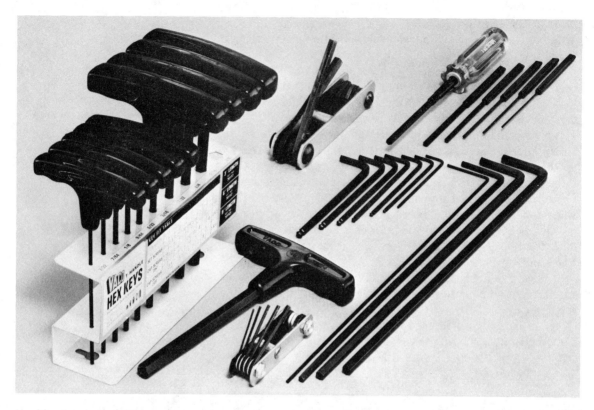

Fig. 6-27. Vaco hex key set.

used for decoration on an engine or in some other spot. Cap nuts are also called acorn nuts, and they are usually chrome plated. I've seen some in brass and other decorative materials.

Elastic locknuts, or aircraft locknuts as they are often called, have a plastic insert that helps keep the nut from vibrating loose. The washer that's built in might also be of other materials, including fiber, composition, and, occasionally, softer metal.

Washers come in three basic styles. Flat washers are used to back up heavy installed parts on softer materials so that the base material isn't torn to shreads around the nut. Split-lock washers are used to help hold a fastener in place against vibration. Such lock washers are usually of spring steel and the ends are meant to dig into both the nut and the work surface to keep things from slipping.

Shakeproof lock washers have teeth and lugs, or one or the other, all around the outside edge. They are designed to grip both the nut and the surface under it. In my experience, these are the least effective, by far, of all the systems for keeping vibration results at bay. Safety wiring is tops and the chemical materials such as Loctite are a close second, assuming properly clean threads. Aircraft locknuts are third. Regular lock washers are a low fourth. The shakeproof washers are dead last and losing ground rapidly.

Keys and pins are items that tend to drop out and get lost. DRI Industries sells kits of various kinds that are nice to have on hand on those days when you realize the pump shaft is turning, but the impeller is not. You note a groove along the shaft that is empty. The impeller is going to spin on that shaft until a key is installed.

Various manufacturers use different shapes and sizes of keys. For straight shafts the most popular two are the square key and the woodruff key. The woodruff key is a piece of arched metal or a half-circle used to keep gears, cams, and pulleys from spinning on a shaft. Fitted and seated correctly, the keys are plenty strong enough to carry quite heavy loads. The variety of available sizes for both types is gratifyingly wide when the time comes to fit them to a particular job.

Round taper pins and *dowel pins* look similar but are quite different. Taper pins are used to locate and position matching parts, but also to secure small pulleys and gears to shafts. The taper is usually ¼ inch per foot. The holes for the pins must be formed with tapered reamers.

If you have to drill out such a pin, you'll probably have to re-ream the hole and go up one size or use a dowel pin in its place (which will not hold as well in most cases).

Dowel pins are useful for positioning and aligning units and pieces of many assemblies. They usually have at least one end chamfered. This is an easily identifiable feature so that taper pins can be readily distinguished from dowel pins. Otherwise, the taper is so slight they could get mixed up. The pin is usually a thousandth or so larger than the hole in which it is to be driven.

Cotter keys or *pins* are used to secure screws, nuts, bolts, and other pins. They're probably the most familiar of the keys and pins because their use is very extensive. Most for general uses are made of low-carbon steel (otherwise known as mild steel), but some are made of stainless steel or other materials as the job requires. Cotter keys come in even prong and uneven prong styles, and the length is measured to the end of the shortest prong.

Cotter keys should fit snugly in their holes with little or no shift to the side. The prong extending above the bolt end should not go beyond the bolt diameter and the end of the bent prong should end before it reaches back down to the washer. Prongs can readily be cut to size with side-cutting pliers.

If you prefer, the prongs can be bent to the sides, along the flats of the nuts. No matter how the prongs are bent, try to avoid getting very sharp radii in the bends. Such sharp bends make the prongs more liable to breaking off under stress. Use needle-nose pliers, not side (or diagonal or dyke) cutters, to make the bends. If necessary, you can use a hammer to get the final bend. Soft-faced hammers are recommended, but my plastic, wood, rawhide, and rubber

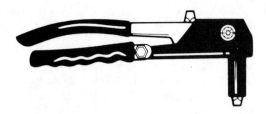

Fig. 6-28. POP Rivetool. Courtesy of Bostik.

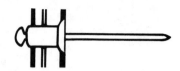

Fig. 6-32. Countersunk rivet. Courtesy of Bostik.

Fig. 6-29. Open end rivet. Courtesy of Bostik.

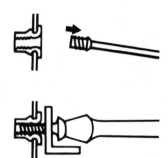

Fig. 6-33. Threaded POP rivet. Courtesy of Bostik.

Fig. 6-30. Closed end POP rivet. Courtesy of Bostik.

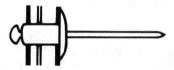

Fig. 6-31. Large flange POP rivet. Courtesy of Bostik.

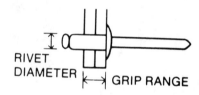

Fig. 6-34. Rivet sizing. Courtesy of Bostik.

mallets just get marked up so I use a light ball peen with gentle taps.

Other forms of threaded fasteners exist for many special-purpose needs. See Figs. 6-24 through 6-27. Turnbuckles are excellent items for tightening things. They also serve well on doors or gates to draw things square again. Turnbuckles consist of a double-ended sleeve with one end threaded for right-hand and one threaded for left-hand threaded rods. Short turnbuckles have screw-eye rods. Longer styles made specifically for forming doors and such will have longer rods to reach from corner-to-corner.

U-bolts, eye bolts, yoke bolts, hook bolts, and a number of other rod-like materials are threaded to accept nuts and can be fastened easily to allow you to hang things.

RIVETS

Rivets (Figs. 6-28 through 6-36) are permanent fasteners that can be driven hot or cold. Hot-driven rivets require tools that are quite expensive. The development some years ago of the POP rivet reduced the extreme costs of riveting. Standard rivets are generally driven hot and require a lot of pressure and a bucking bar in back to form the rivet into final shape. POP rivets require only a riveting gun and a drill but to fit the proper size hole.

Standard rivets are generally made of the same material as that being joined so that galvanic corrosion will not be a major problem. Copper and other soft metal rivets do not require heat. Even steel rivets might not require heat if a source of compressed air of enough pressure is available.

In any case, the procedure is similar up to the forming of the rivet. A hole is centered, punched, and drilled in the clamped materials to be riveted and the burrs are removed. You then use a rivet equal to the thicknesses of the two materials being joined, plus about 1½ times the rivet diameter, that will be just a hair smaller than the drilled hole.

The rivet is then inserted in the drilled hole, and the materials are turned upside down on an anvil (so the rivet head is on the anvil) or a special anvil device is used where pieces are too large to move. The deep hole of the rivet set is placed over the rivet shank and struck with a hammer. Remove the set, turn the material over, and lightly strike the head of the rivet to dish it.

Fig. 6-35. Installing POP rivets. Courtesy of Bostik.

Finally, the heading die is placed on the shank and struck again

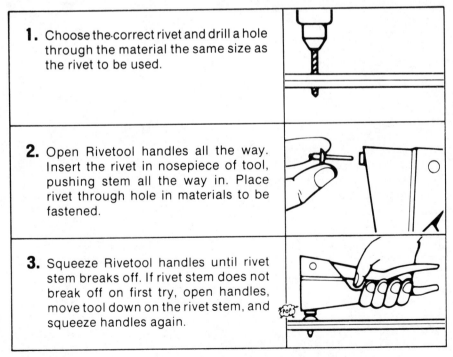

1. Choose the correct rivet and drill a hole through the material the same size as the rivet to be used.

2. Open Rivetool handles all the way. Insert the rivet in nosepiece of tool, pushing stem all the way in. Place rivet through hole in materials to be fastened.

3. Squeeze Rivetool handles until rivet stem breaks off. If rivet stem does not break off on first try, open handles, move tool down on the rivet stem, and squeeze handles again.

NOTE: To remove rivet stem, open handles fully and shake gun.

Rivet Sizes	Rivet Dia. in Inches	Grip Range in Inches	Description	CARDED				BOXED			
				Cat. No.	Rivets per Card	Cards per Case	Case Weight	Catalog Number	Rivets per Box	Boxes per Case	Case Weight
Short	1/8	1/8	Short Steel	S42	25	100	11 lbs. 2 oz.	S42-100 S42-500	100 500	10 10	3 lbs 2 oz 16 lbs 0 oz
Medium	1/8	1/4	Medium Steel	S44	20	100	10 lbs. 7 oz.	S44-100 S44-500	100 500	10 10	3 lbs 11 oz 17 lbs 0 oz
Long	1/8	1/2	Long Steel	S48	15	100	10 lbs. 6 oz.	S48-100 S48-500	100 500	10 10	4 lbs 9 oz 21 lbs 13 oz
Short	1/8	1/8	Short Aluminum	A42	25	100	5 lbs. 14 oz.	A42-100 A42-500	100 500	10 10	1 lb 10 oz 6 lbs 8 oz
Medium	1/8	1/4	Med. Aluminum	A44	20	100	5 lbs. 9 oz.	A44-100 A44-500	100 500	10 10	1 lb 10 oz 7 lbs 3 oz
Long	1/8	1/2	Long Aluminum	A48	15	100	5 lbs. 8 oz.	A48-100 A48-500	100 500	10 10	1 lb 15 oz 8 lbs 5 oz
Short	5/32	1/8	Short Aluminum	A52	20	100	6 lbs. 10 oz.	A52-100	100	10	2 lbs 5 oz
Medium	5/32	1/4	Med. Aluminum	A54	20	100	6 lbs. 14 oz.	A54-100	100	10	2 lbs 8 oz
Long	5/32	1/2	Long Aluminum	A58	15	100	6 lbs. 10 oz.	A58-50	50	10	1 lb 7 oz
Short	3/16	1/8	Short Aluminum	A62	18	100	8 lbs. 6 oz.	A62-50 A62-250	50 250	10 10	2 lbs 0 oz 8 lbs 6 oz
Medium	3/16	1/4	Med. Aluminum	A64	15	100	7 lbs. 12 oz.	A64-50 A64-250	50 250	10 10	2 lbs 0 oz 8 lbs 13 oz
Long	3/16	1/2	Long Aluminum	A68	12	100	7 lbs. 11 oz.	A68-50 A68-250	50 250	10 10	2 lbs 7 oz 10 lbs 8 oz
Ex Lg	3/16	5/8	Extra Long Aluminum	A610	10	100	7 lbs. 0 oz.	A610-50 A610-250	50 250	10 10	2 lbs 7 oz 10 lbs 11 oz
Short	1/8	1/8	Short Steel Large Flange	S42LF	25	100	11 lbs. 6 oz.	S42LF-100	100	10	4 lbs 2 oz
Short	1/8	1/8	Short Aluminum Countersunk	A42CS	25	100	11 lbs. 4 oz.	A42CS-100	100	10	3 lbs 2 oz
Med-Long	3/16	3/8	Medium Long Steel XL Flange	N/A	—	—	—	S66LF-25 S66LF-100	25 100	10 10	3 lbs 6 oz 12 lbs 3 oz
Short	1/8	1/8	Short Aluminum White Painted	A42W	25	100	5 lbs. 14 oz.	A42W-100 A42W-500	100 500	10 10	1 lb 10 oz 6 lbs 8 oz
Short	1/8	1/8	Short Aluminum Brown Painted	A42B	25	100	5 lbs. 14 oz.	A42B-100 A42B-500	100 500	10 10	1 lb 10 oz 6 lbs 8 oz
Short	1/8	1/8	Short Aluminum Closed End	A42CE	25	100	5 lbs. 12 oz.	A42CE-100	100	10	1 lb 10 oz
Medium	3/16	1/4	Med. Aluminum Closed End	A64CE	15	100	7 lbs. 5 oz.	A64CE-50	50	10	2 lbs 1 oz
Short	1/8	1/8	Short Copper	C42	25	100	11 lbs. 10 oz.	C42-100	100	10	3 lbs 14 oz
	1/4	1/16	Aluminum Threaded POP® Rivets	ATPR-8	8	100	4 lbs. 8 oz.	ATPR-100	100	10	2 lbs 6 oz

to form a rivet head. If the original head is not to be flattened, it must be placed in a cupped depression on a second rivet set. You can do the whole job without the rivet set but you get a somewhat sloppier job. All in all, it's a lot of work, most of it unnecessary, and always forces you to have access to both sides of the materials being joined.

Molly's POP rivets are another story. These squeeze-style rivets—you squeeze the rivet gun handle—are also known as *blind rivets* because access to only one side of the work is needed. The rivets used are two part. One major disadvantage used to be the fact that the mandrel, corresponding to the shank in standard rivets, used to be pulled through and dropped out so that the final job had a hole through the rivet center. That is still the case with standard POP rivets, but need not be as solid rivets are now available at a bit extra cost.

The POP rivet requires you to drill a hole to match the listed rivet size. Most rivet styles are available in diameters from ⅛ to ¼ inch and lengths suitable for work up to ½ inch thick. Most commonly, POP rivets are available in steel and aluminum. You'll also find some in copper and one or two other materials. Colors are made to match white and brown siding, guttering, and so on.

Aluminum rivets are best for lightweight jobs. They should always be used when the work is to be aluminum joined to aluminum. Steel rivets take a heavier load and copper rivets are best in copper projects. Copper and aluminum will not rust. Most copper rivets will discolor and eventually turn the characteristic green of the major element involved.

Select either closed-end or open-end rivets. Use closed-end rivets when a seal, whether gas or liquid tight, is required. A properly done rivet, using a closed-end rivet, will be gas tight, liquid tight, and will do a fine job of sealing small pinholes. The open-end rivet is a bit less expensive but it will not provide a seal because the mandrel pulls out and leaves a hole. This hole can be sealed with caulk, silicone sealant, solder, or other materials if you find a need for a seal. Generally, if any need for a seal is expected, the closed-end rivet is far better.

My POP Rivetool offers a selection of tips for different rivet sizes. Tip changeover is fast and a small wrench for the purpose slips into the front of one of the rubber or plastic handgrips. The second tip stores on top of the tool, screwed into place.

POP's general rivet supply range starts at ⅛ inch—moving through aluminum, steel, and copper in that range—and then on to 5/32-inch aluminum, 3/16-inch, and ¼-inch aluminum.

Others are available, from POP, Townsend, and other companies. The grip range extends from 1/16 inch through ⅝ of an inch. Back-up plates are available for installation where there might be a

Fig. 6-36. Left: POP rivet selection. Courtesy of Bostik.

211

lot of stress or where one material is appreciably softer than another to which it is joined.

POP also makes a variety pack in the form of a wheel with six different aluminum and steel rivets. This selection includes three sizes of steel rivets and three of aluminum, all in ⅛-inch diameter, but in ⅛-, ¼-, and ½-inch grip ranges. Those are the most popular and they make a fine basic selection. There are some 120 rivets in the pack and it's easily stored.

The installation of POP-style rivets requires little time or effort compared to installation of standard rivets. The hole is drilled through the pieces to be joined and the rivet is inserted into the rivet gun with handles opened wide. Set the pointed end of the mandrel as far into the tip as it will go and put the bulbed head of the rivet into the holes. Make sure the rivet head base is against the materials being joined. Squeeze the handles together. That's it. The job's done. Open the handles, shake out the mandrel, and go on to the next rivet if you need one.

Metal back-up plates are used in soft materials or where stresses will be great. Allow for their thickness when determining rivet grip range.

If you have heavy-duty riveting to do, you'll want to investigate the rental of long-handled rivet guns used on commercial jobs. Check into rivet size availability or use standard rivets.

When you must join metals and can't use rivets, nuts, bolts, screws, and such materials, you've reached the time to start considering one of the three heat processes for joining metals. Soft soldering, brazing, and fusion welding all offer advantages over other methods. Some disadvantages that must be considered start with tool cost in the brazing and welding fields and joint weakness in soldering.

Chapter 7

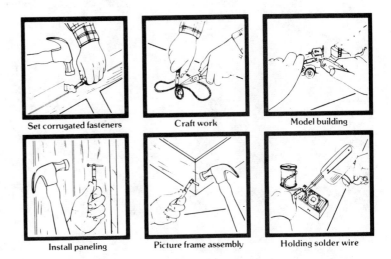

Set corrugated fasteners Craft work Model building

Install paneling Picture frame assembly Holding solder wire

Heat Joining

There are three primary processes involved in heat joining metals, with some suitable for one thing, and some for others. The strongest is fusion welding of similar metals (dissimilar metals cannot be fusion welded). Brazing or braze welding is in some circumstances preferable for strength over fusion welding (cast iron is a good example). Brazing also serves to join dissimilar metals with great strength.

Hard or silver soldering is at the top end of the soldering process—or the bottom of the brazing process—and offers features of both. Soft soldering, using tin-lead solders, is the third primary method of heat joining metals. Softer solders do not offer good joint strength. They are suitable for a number of uses, including electrical contact work—soldering wires to terminals, for example—a sweat soldering copper pipe for plumbing and other needs. See Figs. 7-1 through 7-24.

Fusion welding is a fairly complex process that requires a modestly large tool outlay, no matter what process you intend using. Arc welding is fine for many purposes and generally tends to produce a stronger weld than does gas welding in many situations. Gas welding is more easily portable. Less expensively portable could be accurate because the gas-driven generators to run arc welders increase the price of the arc welders by about a factor of nine. A $250 arc welder will do virtually every job you or I will ever

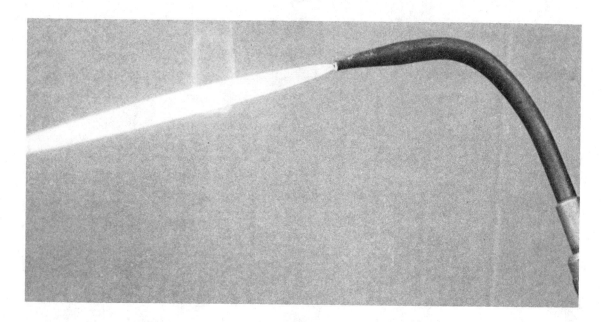

need, but the generator to drive it in the field—plus a trailer to haul the thing and a trailer hook-up for car or pick-up—can readily add $2,000 to the arc welder cost. A gas-welding outfit, with cart and a year's cylinder rental, will usually remain well under $500.

ToteWeld outfits for gas welding are made by Airco. The package, including cylinders, usually costs in the vicinity of $225, with a carrying holder. The smaller cylinders, and the version that uses MAPP gas instead of acetylene, mean that it's far simpler to cart things around to do smaller welding jobs. I've used the ToteWeld to repair some decently large farm equipment. Welding

Fig. 7-1. Torch being adjusted for welding.

Fig. 7-2. Large soldering iron. Courtesy of Weller/Cooper Group.

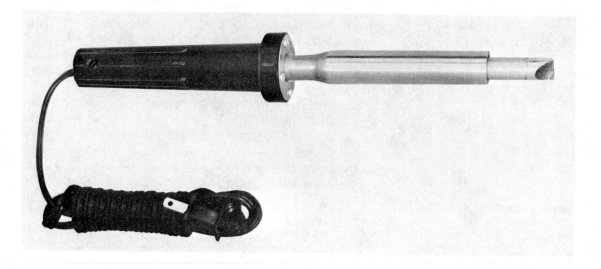

Fig. 7-3. Adjusting the oxygen on a
ToteWeld unit.

the material took a great deal of care and more passes than I'd normally care to make.

For around-the-home use, that an arc welder is the best way to go and the least expensive choice. If any sort of portability is needed, go with gas equipment, with a ToteWeld unit if your needs are modest, and with a full-sized outfit—always including a large wheeled cart—if you expect to have to do heavy cutting or welding.

If you do a lot of metal forming or metal cutting, the gas equipment is a far better bet than the arc-welding gear. While a carbon arc torch works decently with metal cutting, the oxy-acetylene cutting torch is far neater and more efficient. The various heating torches available for tip replacement on gas torches make preheating of areas to be welded or brazed far easier. This is also true for areas that must be heated intensely for bending operations and other forming needs.

Very lightweight gas welding outfits from various companies such as Bernzomatic can be handy, but they are generally most useful only on work with sheet metals. The price differential is great. Virtually all of the super lightweights sell for under $55. If

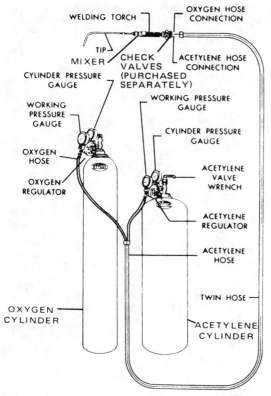

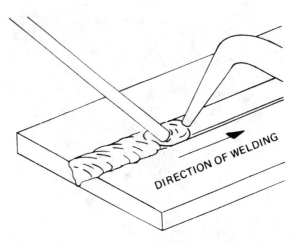

Fig. 7-6. Backhand welding. Courtesy of Airco Welding Products.

Fig. 7-4. Oxy acetylene set up. Courtesy of Airco Welding Products.

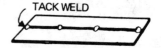

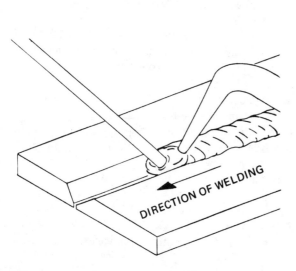

Fig. 7-5. Forehand welding. Courtesy of Airco Welding Products.

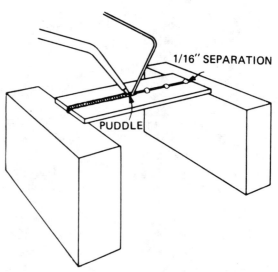

Fig. 7-7. Basic welding. Courtesy of Airco Welding Products.

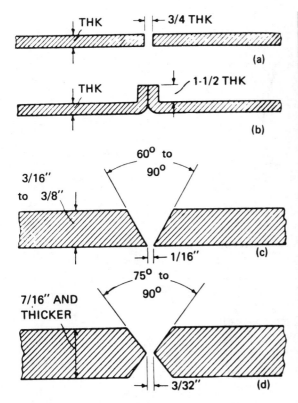

THK → ← 3/4 THK

(a)

THK ↑ 1-1/2 THK

(b)

3/16" to 3/8"

60° to 90°

← 1/16" (c)

7/16" AND THICKER

75° to 90°

← 3/32" (d)

Fig. 7-8. Jointer preparation. Courtesy of Airco Welding Products.

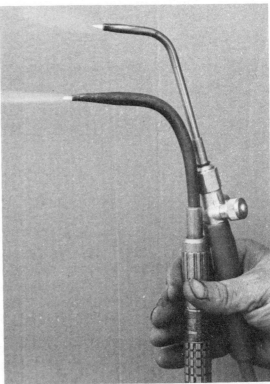

Fig. 7-9. Oxyacetylene flame (bottom) compared to a MAPP flame.

your needs are limited to sheet metals and light tubing, such a tool would be advisable.

Arc welders are available in a wide range of sizes. Some are made to use with 20-ampere, 115-volt circuits. Most require a 50- or 60-ampere at 240 volts. Some large units require 200 to 300 amperes. The arc goes from cold to about 6000 degrees Fahrenheit almost instantly. The melting of the parent metals and the electrode material is quick and there is little chance for the formation of oxides. Included oxides, as they're called, weaken welded joints.

Light-duty arc welders will generally do heavier welding than will light-duty gas welders. The heavy-duty welders will out perform heavy-duty gas welders.

Fusion welding requires some practice in forming tack welds and in running basic beads. You'll want to get that practice in before trying anything complex. One of my favorite exercises used to be to cut open a couple of coffee cans, afterwards heating them enough to burn off the tin coating. Then I tack welded the cans together, cut them apart again and welded with rod. You can also run practice beads on the surfaces of the cans.

217

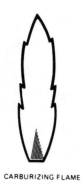

CARBURIZING FLAME   NEUTRAL FLAME   OXIDIZING FLAME

Fig. 7-10. Oxy-MAPP flames. Courtesy of Airco Welding Products.

If you end up able to do such work consistently without warping the very light sheet metal in the cans—or warping it very little—then you'll be able to go on to heavier work with ease. Welding of light materials is usually more difficult than welding of heavy materials. The light materials warp far more easily. They also have a strong tendency to burn through.

While heavy-metal joints might require some grinding to get the proper joint shape for greatest strength, burn through at the weld is less of a problem, by far, than is warping. Of course, heating metals to the 2000-degree-Fahrenheit range is liable to cause all sorts of changes, depending on the metals involved. Stainless steel

Fig. 7-11. Braze joints. Courtesy of Airco Welding Products.

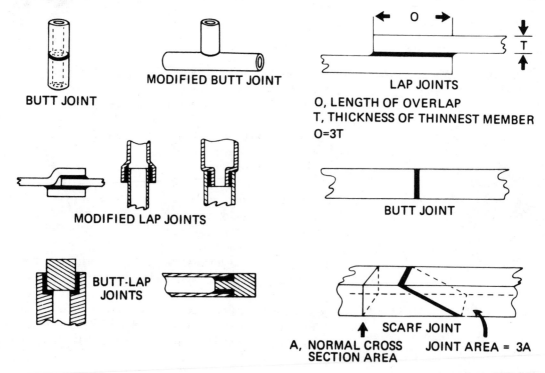

BUTT JOINT

MODIFIED BUTT JOINT

LAP JOINTS
O, LENGTH OF OVERLAP
T, THICKNESS OF THINNEST MEMBER
O=3T

MODIFIED LAP JOINTS

BUTT JOINT

BUTT-LAP JOINTS

SCARF JOINT
A, NORMAL CROSS SECTION AREA   JOINT AREA = 3A

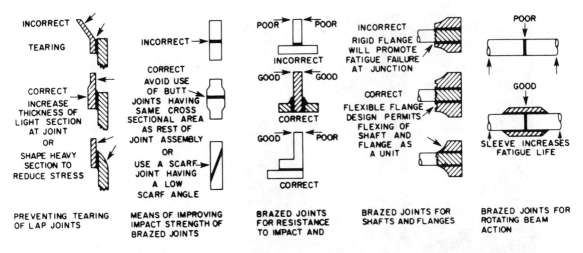

INCORRECT

TEARING

CORRECT
INCREASE
THICKNESS OF
LIGHT SECTION
AT JOINT
OR
SHAPE HEAVY
SECTION TO
REDUCE STRESS

PREVENTING TEARING
OF LAP JOINTS

INCORRECT →

CORRECT
AVOID USE
OF BUTT
JOINTS HAVING
SAME CROSS
SECTIONAL AREA
AS REST OF
JOINT ASSEMBLY
OR
USE A SCARF
JOINT HAVING
A LOW
SCARF ANGLE

MEANS OF IMPROVING
IMPACT STRENGTH OF
BRAZED JOINTS

POOR →   ← POOR

INCORRECT

GOOD →   ← GOOD

CORRECT

GOOD →   ← POOR

CORRECT

BRAZED JOINTS
FOR RESISTANCE
TO IMPACT AND

INCORRECT
RIGID FLANGE
WILL PROMOTE
FATIGUE FAILURE
AT JUNCTION

CORRECT
FLEXIBLE FLANGE
DESIGN PERMITS
FLEXING OF
SHAFT AND
FLANGE AS
A UNIT

BRAZED JOINTS FOR
SHAFTS AND FLANGES

POOR

GOOD

SLEEVE INCREASES
FATIGUE LIFE

BRAZED JOINTS FOR
ROTATING BEAM
ACTION

Fig. 7-12. Joint designs for brazing. Courtesy of Airco Welding Products.

Fig. 7-13. Imperfectly heated brazing material.

is easily ruined by too much heat. It is usually hard soldered, silver soldered, or brazed instead of fusion welded.

When great strength is required and vibration resistance must be total, welding pieces into a single part is just about the only sensible way to go.

FUSION WELDING

Fusion welding is a process used to turn similar metals into a single piece of a shape and design to do a particular job. It's also used to make repairs on pieces that must be solid.

For gas welding—whether oxyacetylene or oxy-MAPP—one of two torch positions are used: forehand welding has the welding rod in front of the torch tip; backhand welding finds the welding rod

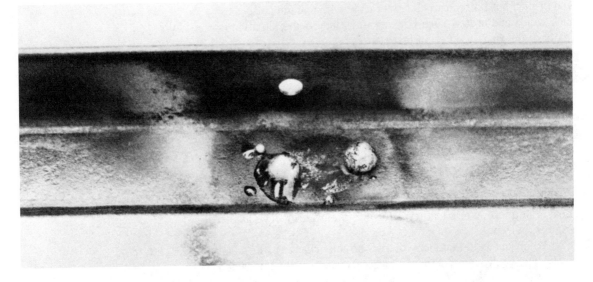

Fig. 7-14. Light Weller soldering tools.

following the torch tip. Because backhand welding results in smaller puddles of molten material, it is preferred for many jobs such as welding vertical materials and overhead welding because there is less chance of the smaller puddle running. Forehand welding is a bit easier to learn to do; with practice both will be of value to you.

Tack welding, with gas or arc welders, is done without the use of a filler rod or electrode. With a torch, set the work so that you have easy access to it, depending on whether you're right-handed or left-handed. Light the torch and get the correct, neutral flame. Move the flame across the work in small circles or back-and-forth along a ¼-inch line at the seam. You can also use a zigzag pattern if you prefer, but I'd recommend practicing with a single pattern for tack welds and pretty much sticking to it, unless the situation or material demands a different process.

With mild steel, you'll readily note color changes as the metal heats up. As the metal begins to turn white from yellow, it will turn molten. At this time, the torch tip should be advancing the flame to a spot that's not quite molten. The puddle is allowed to cool and set. Keep torch speed slow so you don't advance too far from the puddle. Too quick movement means the weld doesn't penetrate well and won't be strong enough. Too slow movement will see burn through at the weld root (in thicker metals).

Other metals are not as easily welded as steel. Your first attempts should be kept to steel, and mild steel (low carbon) at that. Leave the more exotic materials for the time when you've gained some experience.

When you weld aluminum, there is no color change in the metal as it heats (you've also got to worry a great deal more about included oxides). All of a sudden you've got a sagging puddle of aluminum that is still just about the same color as the base material that hasn't been heated!

Your best bet is to keep some scrap aluminum of different sizes on hand. Make practice welds in material the same alloy and thickness as the metal to be finally joined. Keep a rough check on weld-forming times. Aluminum is most often ruined by too much heat in small sections. Heat should be applied for little more than five or six seconds on lighter materials.

Welded joints are allowed to cool naturally in most cases. Dipping in water has been often used to speed things up. I've not had much luck with the water-dip method because it seems to add too much to the metal stresses during cooling. Metal or weld failure becomes a problem.

If you're going for the extra strength welding provides, it somehow seems illogical to take chances on losing that gain just to save 15 minutes of cool-down time. If you want to work with annealing, softening and hardening metals, I recommend that you find a book totally on the subject such as *Do Your own Professional Welding,* TAB Book No. 1384, or *The Welder's Bible,* TAB Book No. 1244.

Fig. 7-15. Comparison of soldering irons and gun. Courtesy of Weller/Cooper Group.

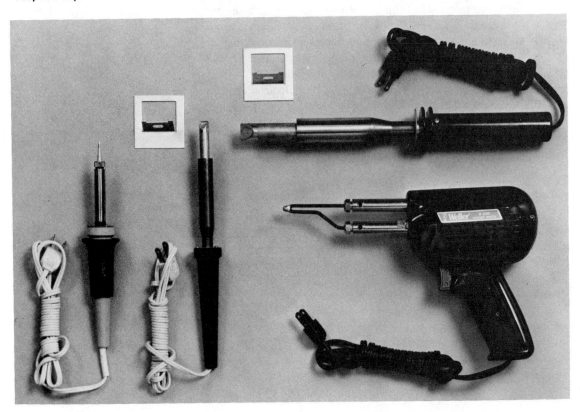

With *arc welding*, you use an electrical source for heat. One line is grounded to the work and the other with an electrode or carbon arc torch. Brazing, tack welding, and such nonelectrode operations are performed with the carbon-arc torch. My preference has always been for the gas welder when torch use is necessary.

Arc welders for home use don't exceed 300 amperes capacity. Make sure the one you get has at least a 60-percent duty cycle. A 100-percent duty cycle is best, but expect to pay extra for a machine that will operate at top output on a continuous basis. That is not often needed for home and farm welding. The smaller the welder you buy, the closer to 100 percent of the cycle you're likely to need should heavy work crop up.

The carbon-arc torch is a single accessory and an inexpensive one for the arc welder. From that point, you are working with machines that can do more jobs more easily, and you're going to have to pay for the extra capacity. The ac machines are generally the least expensive. The ac/dc and dc are the most useful because they can be added to using tungsten inert gas welding units. The TIG welding unit will allow for easier, stronger welding of more exotic materials.

The TIG unit will have an arc stabilizer—this works with or without the accompanying gas shield—so that a long arc is far more easily maintained. The result is a stronger, neater weld. Gas shields allow welding of odd metals by excluding air, thus cutting way back on oxide formation, and providing a clean, strong weld.

TIG welding uses a tungsten electrode. Any filler metal needed, which would normally be provided by the electrode but is not with this process, is provided either by a separate electrode or rod or by a chunk of the original metal cut from a scrap piece. This allows an exact materials match in situations where electrodes are not available in close enough alloys to make strong welding possible.

The inert gas used—in a large cylinder and fed at a metered rate to the weld site—is usually argon.

The melted electrode, or melting electrode, in standard arc welding is the filler metal and source of heat combined. It passes electrical current to the work attached by the ground cord to the welder. It slowly melts to deposit any required material in the weld. The shield needed to cut back on oxide formation is made when the chemical coatings on the electrode heat and melt off the electrode. This shield of fluxing chemicals is a constant in many forms of welding—whether it's arc welding, gas welding, brazing, or even lowly soft soldering.

Oxide formation is the biggest problem with any form of heat fastening because the increased heat speeds up the chemical actions that form oxides. The more thoroughly air is kept from any joint

Fig. 7-16. Right: Soldering gun kit. Courtesy of Weller/Cooper Group.

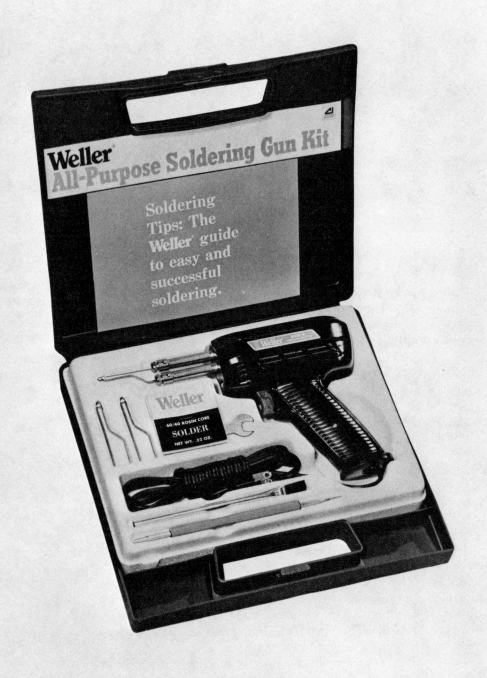

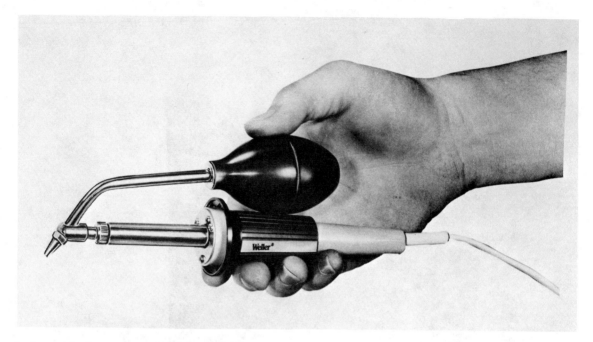

being made the stronger the joint will be.

Both arc and oxy-fuel gas welding are handy skills to have for any crafts person. If all you ever need to do is tack weld a piece or two of metal, that's reasonably true. But if that's the case, don't waste good money on welding gear. Have someone else do it.

Welding requires some hours of practice. You must learn all that's possible about the tools involved, the electrodes and rods, American Welding Society specifications for tools and rods and electrodes and fluxes, and the reactions of various metals to varied amounts of heat. Read this section and examine the illustrations as many times as you feel necessary. Then practice a bit, and then read some more complete books. Practice some more, and then take a local course in welding. After that, you might begin to see how much is involved. For simple one-time jobs, you'll be way over-qualified. You'll also be able to do complete projects that now seem impossible or projects that won't even come to mind.

Safety is always a most important part of working with high-amperage electricity or high-pressure gasses. On that basis, I would recommend following all installation details to the last letter for arc welders, and going with oxy-MAPP gas welding materials. MAPP is a stabilized form of acetylene and comes very close to producing just as much heat. It's often more readily available than is acetylene because it comes in small cylinders.

Check your local telephone directory for welding supply distributors. You'll save about 25 percent on the cost of each cylinder. The same holds true for welding rod and electrode and other materials. You pay through the nose in the standard retail establishment, and the guy running the welding supply house will be just as happy to have your business. Most materials come in small enough units to make the purchase by amateurs reasonable.

BRAZING

Brazing uses a brass alloy filler (all brass is an alloy anyway) to form joints with high heat, far stronger than soldering can produce. In some cases, the joint is as strong as you would find with fusion welding. It is of particular use where the extra heat of fusion welding might destroy certain qualities in a metal or where the metal has peculiarities that don't really allow for good fusion welding. A case in point on the last part is the brazing of cast iron. Whether you need to repair a tractor manifold or an engine block, cast-iron material has quite a lot of included carbon. That carbon makes fusion welds brittle and subject to easy breakage. There is a great loss of strength when heat is applied. Many cast-iron sections, as in engine blocks, are so large as to preclude even heating of the part needing repair without the use of very large preheating torches in addition to the basic welding torch. Virtually all brazing is done

Fig. 7-18. Note the thin, small tip on the top soldering iron. It is suitable for electronic work. Courtesy of Weller/Cooper Group.

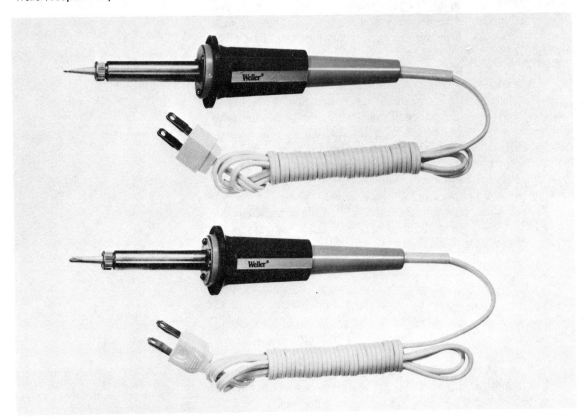

with oxyacetylene or oxy-MAPP torches. It can be done with a carbon-arc torch, but gas equipment is more efficient and usually a lot faster and easier to control. You use the same torch, and often the same size torch tip, as for gas welding.

The joint is formed as required, with or without grinding to shape, and you then preheat the cast iron until flux applied flows smoothly and freely. The brazing rod is then heated and applied with the same torch movements as in fusion welding. With heavy castings, a slow cool down is best. A heatproof covering of some sort is needed. A heat application, less and less each time, every 15 minutes or so of about one application for every ½ inch of casting thickness is handy. The cooldown heatings should just warm the material, without bringing it anywhere near the temperature needed to melt down the brass rod, and each heating should be a bit shorter than the previous one.

In general, brazing is a simple process; the same joint designs are used as for fusion welding. It is also an excellent spot to begin one's "career" in welding. It is easier and a bit faster, though most of the process is identical and transferable to later fusion welding jobs.

Joint designs are similar to those used for soft soldering in overall design and joint-clearance needs. A bit less mechanical support of the joint is needed because the higher heat levels give a stronger bond (as does the stronger bonding metal). Dissimilar metals can be joined, as with brazing, and there are a number of uses for silver soldering that give it wide-spread use around the shop.

Most stainless steel has a strong tendency to lose many of its valuable characteristics—including its non-rusting properties—when excessive heat is applied. Thus silver soldering becomes the method of choice when assembling or repairing stainless items. Silver soldering is also useful for making and repairing jewelry. A soft soldered piece of enamel-surface jewelry will drop its coat quickly; silver soldering will hold. At least it will hold until the melting point of the bonding alloy is reached.

Surfaces are prepared and cleaned with the flux brushed on the parts to be joined. As in all forms of soldering, do *not* get flux on those areas where you don't want to have solder showing. When the flux bubbles and then smooths out again, the joint is ready for flux application.

Some fluxes are set to change color to give more obvious indications of readiness. If the solder flows well once it's touched to the heated surface, that isn't really necessary. If the solder balls up, the joint is too cool. If it goes watery and drips off, the joint is too hot.

Silver soldering can be carried out with propane and MAPP gas torches in most cases. MAPP is preferable because of the hotter flame, but for small jobs propane will work quite well.

Fig. 7-19. Right: A soldering iron stand is an aid to both safety and efficiency during work. Courtesy of Weller/Cooper Group.

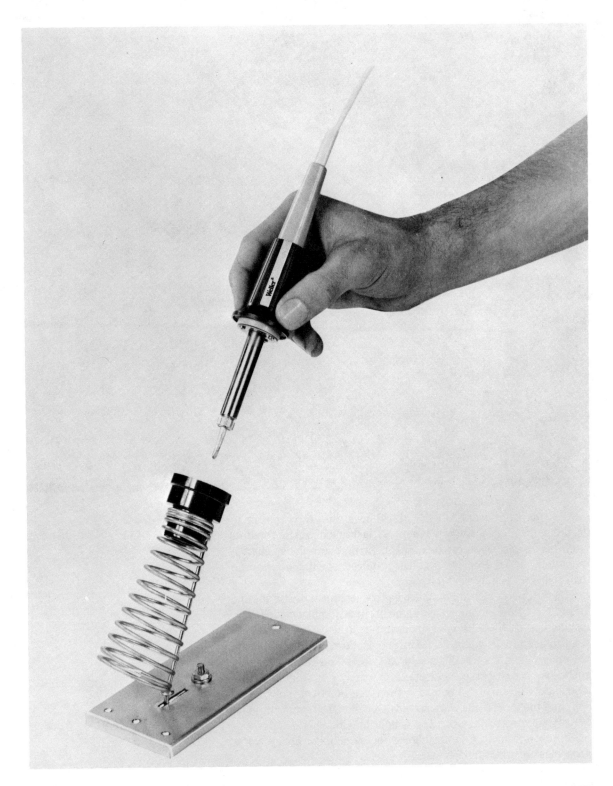

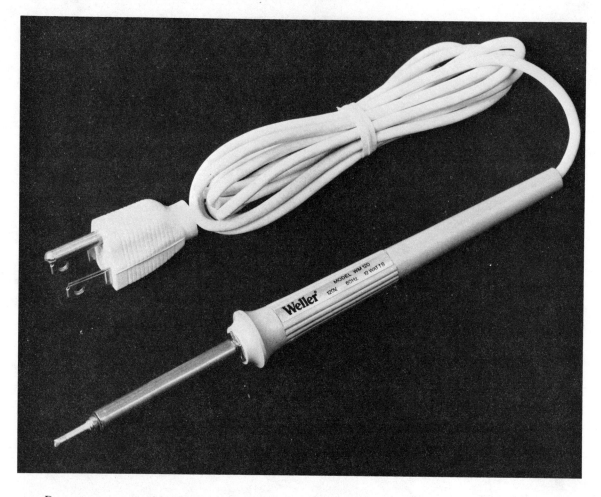

Do *not* ever use a MAPP gas cylinder on a torch meant to handle propane. Propane cylinders work just fine in MAPP torches, often better than with torches meant for propane alone, but the reverse is never true. MAPP gas requires slightly heavier tip materials and a different tip design.

New torch designs that change gas movement at the tip are fine for producing hotter flames, and some propane torches made in this manner approach MAPP gas temperatures for use. You still do not want to use MAPP with them. The propane torch offers one great advantage over the MAPP gas torch. Cost. The torch generally costs less at the outset, and the fuel, even at a welding supply distributor, is about 40 percent of the price of MAPP.

Suit the torch to the work and you'll do fine. You can, with some MAPP gas torches, even carry out light brazing jobs. An oxy-fuel gas torch is usually a better choice for medium to heavy work, regardless of what the ads say.

Fig. 7-20. Weller's 12-watt gun is suitable for the lightest soldering work.

SOFT SOLDERING

Soft soldering uses are wide around most any home or shop. The tools used can vary from soldering guns to soldering irons to propane and MAPP gas torches. The tool selected should match the job to be done.

Electrical and electronic soldering requires a number of considerations that sweat soldering of open pipe runs does not. First, the flux for the solder must be of a type that doesn't cause corrosion. Acid fluxes cause corrosion to a degree that's no real problem in sweat soldering. It just looks kind of ugly as the copper pipe goes kind of green. Do that with electrical work and you have a high resistance area that will mess up the flow of electricity and that may cause fires. Thus a noncorrosive flux—a rosin flux—is needed. Whatever flux you're using for any job, don't get it on your hands any more than is essential. And don't get any of it in your eyes at any time.

Small soldering irons are widely available today; a number of good brands are on the market. An Ungar model has interchangeable heating elements. The Weller models, with their interchangeable screwdriver style tips, are handy for a number of jobs the Ungar

Fig. 7-21. Emery cloth does a fine job of mechanical cleaning of copper pipe.

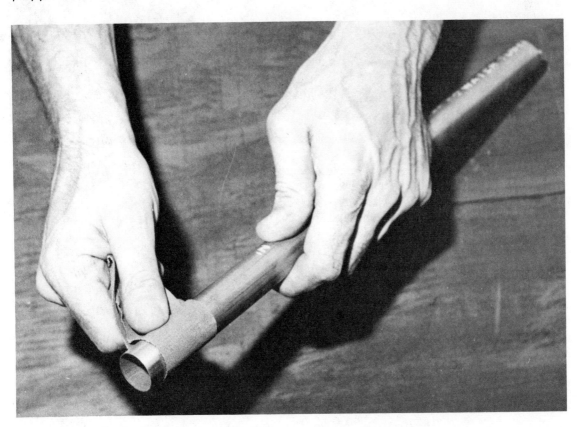

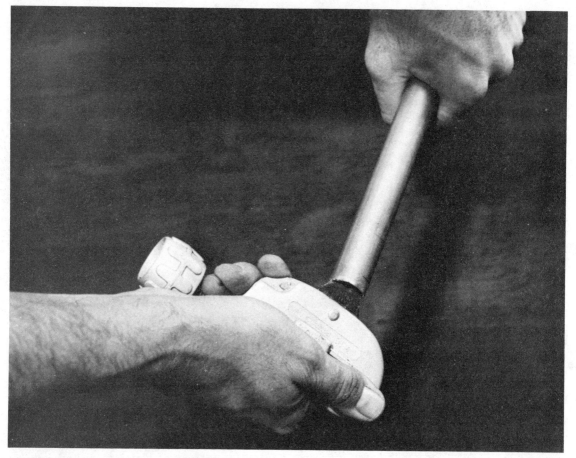

won't handle because of size. Weller also makes other irons and soldering guns.

Fig. 7-22. Make sure you get interior pipe burrs out with a reamer so water flow is not hindered.

The Weller models with small tips give a temperature around 600 degrees Fahrenheit (⅛ inch tip, CT6C6 catalog number). The tip sizes go up to ⅜ inch. Also available are 600-degree models, 700-degree models and 800-degree models. For desoldering uses, I've got a Weller DS40, which is a 40-watt iron, with a holed tip leading to a rubber bulb. Heat the old solder, squeeze the bulb, release and you've got a clean terminal ready for the next lead to be soldered on. Tip temperature on this tool is about 750 degrees, and it works quickly and far better than simple bulbs that require you to heat things up with a separate soldering iron. Various tips are available for the desoldering iron. Sizes range from .038 inches to .090 inches.

Weller makes solder in both rosin and acid-core styles (in small and medium sized containers). This wire solder is very easy to use and store, and it comes in several diameters. The smaller size is used for light jobs and the heavier jobs are carried out with the thicker diameter solders. The tin-lead content is 40-60 in these solders, and the plasticity and flow are excellent for the hobbyist who doesn't want or need to gain great experience in dealing with more difficult to use soldering materials.

When soldering, mechanical cleaning is carried out first (usually) with emery cloth or sandpaper. The cloth is preferable because it tends to drop fewer abrasive particles. Steel wool is particularly handy when cleaning the ends of pipe to be sweat soldered and the interior of pipe joints.

After that, flux is applied—in either liquid or paste form—if the solder is solid. Flux need not be applied if you use a cored solder. I prefer, for sweat soldering, solid-core solder with either a rosin or acid flux in liquid form.

Heat is applied after the iron tip is tinned. Clean with steel wool or a piece of emery cloth while warm, and then apply a light coat of fresh solder. You then apply the solder to the heated part. If the part is hard to reach with the solder, you might find it necessary to apply the solder to the iron tip and to allow it to flow onto the part. This is inefficient and tends to allow for a lot of waste dripping in where it's neither wanted nor needed—and where it may later cause problems.

The solder will form a joint with a smooth, shiny surface. At that time, you remove the tip and let the joint cool down before it is moved at all. Moving the joint while the solder is still liquid will virtually always cause a poor electrical connection.

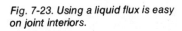

Fig. 7-23. Using a liquid flux is easy on joint interiors.

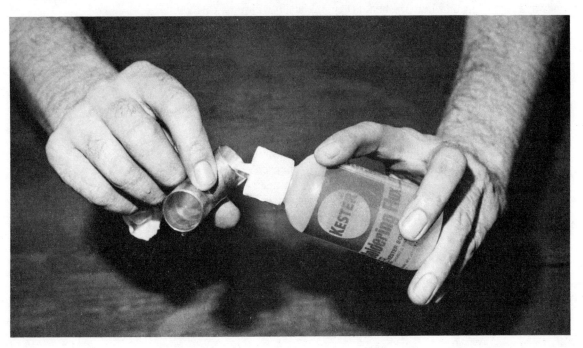

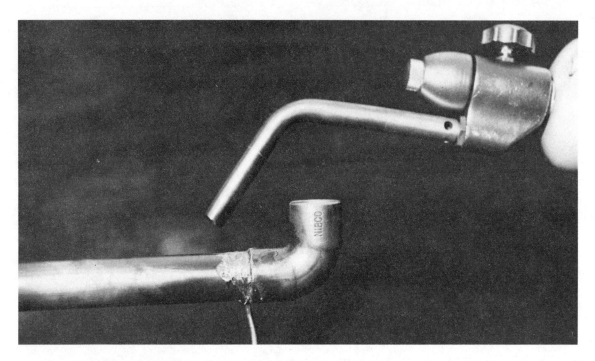

Fig. 7-24. Apply solder on the side opposite heat application.

Sweat soldering of copper pipe for plumbing, or other uses, is a process that depends on capillary action. The solder will flow to the heated and fluxed area and will remain there when the heat is removed, thus sealing the joint. Solder will flow up, down, or sideways depending on where the heat is applied, and it will remain in the cleaned and fluxed areas.

Materials to be sweat soldered are first mechanically cleaned with steel wool or emery cloth, after being tested for fit (a tight joint now means a tight joint later, as the solder is not meant to bridge large gaps). Don't clean things mechanically so hard that things get sloppy at the joint. Clean just enough to create a bit of shine on the copper parts.

Assemble the parts after applying flux just to those portions that are to have solder on them. I prefer a liquid flux, but others swear by paste fluxes. Don't try to do this job with cored solders because they tend to make the work sloppier than it needs to be. The solder will stick to any part that has been cleaned and heated, and flux-cored solders will flux everything they flow across.

Apply heat using a propane or MAPP torch. Depending on the size of the pipe to be joined, you might or might not need the extra capacity of the MAPP torch. Propane is more than sufficient for pipe up to 1½ inches inside diameter and is usually sufficient to 3 inches and more.

Heat is applied to one side of the assembled joint and the solder is touched to the other side of the joint. The joint is ready to be soldered when the flux either changes color or bubbles, depending on its type. The solder will flow to the heat. Keep moving the torch around the pipe until you have a good seal on all sides. Such a seal can be seen when the solder forms all the way around.

Don't move the joint until after the solder has had plenty of time to cool and set. Early movement will break the seal and make for a messy disassembly and reassembly. It is not easy to disassemble and reassemble soldered joints neatly.

Precautions are always necessary when working with heat and especially with open flames. Make sure that no flammable materials are close by, and make sure the torch flame does *not* touch joists or other combustible materials.

Use spun ceramic fireproof materials to prevent such contact if necessary. Use the torches only in ventilated areas. They don't consume a lot of oxygen, but an enclosed space is nowhere to be losing any oxygen. Shut the torch off before laying it down. If this is a problem, ease it by selecting one of the newer torches with a pilot light and a trigger.

Store the gas cylinders with care. Make sure they are not in a closet in a bedroom, for example, and also making sure the temperature stays well under 120 degrees Fahrenheit. If you think you've got a leaking valve on your torch, test it with a liquid soap and water solution just as you would test a welding torch. Welding torches must always be tested when being attached to new hoses, etc.

Either wear a mask or keep away from any fumes the torch might kick out while it's in use. This is especially true if you're heating something that has been galvanized. Zinc fumes will make you very ill or worse. Make sure, if you're soldering a container, that it has either not contained flammable materials—kerosene, gasoline, motor oils, paints, etc.—or that it has been through a proper purging (a complex and often expensive job). All closed containers must be well vented before they are heated.

In general, use care when working with any source of great heat. Take even more care when that great heat comes from a highly compressed gas that is not good for your lungs or the rest of you. When welding, brazing or soldering, follow all manufacturer's safety directions.

Chapter 8

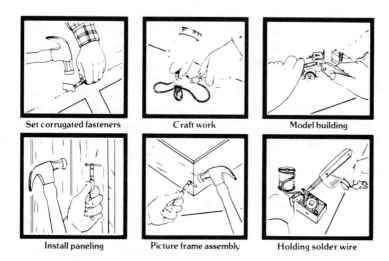

Set corrugated fasteners   Craft work   Model building

Install paneling   Picture frame assembly   Holding solder wire

Flexible Fasteners

In most cases, flexible fasteners such as clamps, tape, and rope (Figs. 8-1 through 8-28) are temporary fasteners meant to do a job of holding things for a specific length of time. Some, such as Velcro tapes, allow the rapid removal and reinstallation, of an item. Some such as rope caulk more or less permanently seal things. Others such as masking tape simply seal out insidious materials.

Transparent tapes such as Scotch tapes, after the original brand name from 3H Company, are available in the old-fashioned slick surface or a newer, less smooth surface that will accept writing and postmarks.

Masking tape is designed for application on a temporary basis. It often is used to fasten sheets down to prevent overspray from reaching large areas during paint spraying jobs or to form a sharp line at an edge during most any kind of painting job. Good masking tape such as those produced by 3M or Manco will unwind easily. In addition, the tape will remain on a surface for 48 to 72 hours *without* causing damage to that surface (damage includes the lifting of paint, among other things).

Masking tape must pull away from the surface easily while still forming a nearly impermeable bond with the surface before it is pulled off. Tape must go on easily and must cut easily when used around trim pieces.

The price must be moderate and the tape must be available in a

Fig. 8-1. A variety of tapes that are handy for around the home.

wide variety of sizes. Widths usually range up to 3 inches and down to ½ inch, or even less in some cases. Don't buy the unbranded junk that supermarkets carry. Most of it won't stick well, will lift paint after about three hours, and won't peel off easily from the roll once the roll gets a bit of age on it.

Duct tape (utility tape) in its most common guise is a plastic or cloth tape. Originally, it was designed to seal vents in heating and cooling systems to prevent loss of air from the ductwork. At first, all duct tape was gray, but today you'll find it in colors as well. Manco makes about 10 tape colors in addition to gray. While Thermwell produces only gray duct tape, their flame-resistant (*not* fireproof) plastic duct tape is a good product.

Duct tape has far more uses than just sealing ducts. I've used duct tape for its original purposes, of course, but also to form nailing edges for sheet plastic storm windows, on screen porches, and

Fig. 8-2. Duct tape.

Frost-King poly weatherstripping tape is made of silver poly coated cotton cloth for maximum flexibility and strength. Self adhesive, it's easy to use for sealing around air conditioners, putting up plastic storm windows or sheeting, use it to wrap around tool handles or seal a garden hose.

Fig. 8-3. Duct tape. Courtesy of Thermwell Products.

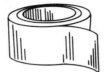

- Double-coated to hold rugs in place
- Eliminates stapling and tacking
- Holds car mats securely in place
- Excellent for carpet tiles

Fig. 8-4. Carpet tape. Courtesy of Manco.

Fig. 8-5. Mounting tabs. Courtesy of Manco.

elsewhere. The plastic tears rapidly if nailed through, even with Frost King's 6 mil weight, but running a double layer of duct tape at a folded over edge before nailing means no tears. If you don't want to nail the sheet up, the duct tape will also serve as an edge sealer—run half on the sheet and half on the window edge.

Cloth and fabric tapes are quite similar to duct tapes. In most cases, the cloth is less durable and the adhesive is not waterproof. I don't like most of them beyond using them to salvage book bindings.

- Double-coated for easy application
- Mount pictures, mirrors, decorations
- Replaces nails
- Cut into convenient tabs

- Meets UPS and Post Office requirements
- Fiberglass-reinforced for extra strength
- Seals packages and bundles securely
- Other household and shop uses

Presumably, such tapes are also useful for fast repairs to plastic upholstery such as that used in most cars and pickups. I've never had much luck. The stuff always lifts off too easily.

Double-faced tape comes in several versions. The heaviest-duty stuff is meant for holding rugs of many kinds in one place. It is also useful at borders of sheet vinyl flooring and in other areas where edge lifting might be a problem.

Scotch mounting squares, from 3M, are lightweight, small squares—about ¾-inch square—of double-faced material. It's thicker than most double-faced tapes, and it is very useful for hanging posters, wall maps, and other such materials. Other companies sell similar products.

A sticky pull-off, reusable adhesive that doesn't have a tape backing, is Bostik's Blu-Tack. This material comes in a small sheet. A piece of the size needed is simply torn off, stuck on the surface, and the material to be hung is stuck to it. Like Scotch brand mounting squares, it's meant for light materials such as balloons, posters, memos, etc. The primary advantage is its reusability. Blu-Tack is nontoxic. It is safe to use for photos, cards, and other things that kids might be handling.

Fig. 8-6. Strapping tape. Courtesy of Manco.

Fig. 8-7. Electrical tape. Courtesy of Manco.

- Tough electrican's grade
- UL-listed
- For electrical splices
- Many other shop and household uses
- In handy cutter-bar dispenser

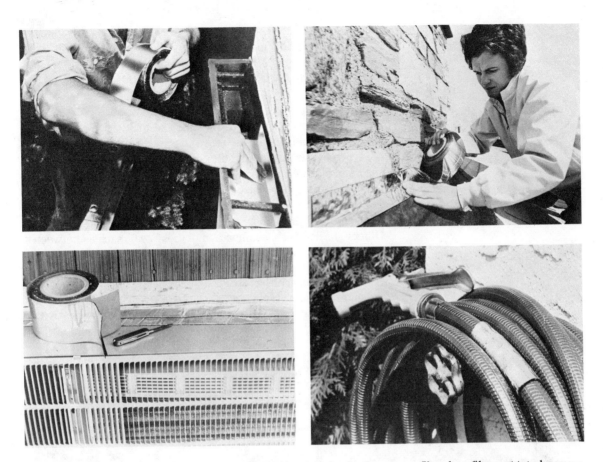

Strapping tape is also known as *fiberglass filament tape* because it has reinforcing strands of fiberglass running through it. It is a superb packaging tape that is exceptionally strong without presenting the problems of opening a package that wire or metal banding presents. Unlike wire, it won't cut into softer packaging materials.

Uses for strapping tape are somewhat wider than advertised in many respects but less than expected in some ways. Because of the glossy surface, it is handy on certain items to go through the Post Office. I've had better than fair luck mending a blown out piece of hose with it.

Electrician's plastic tape seems to be the best answer by far to the old friction and rubber tape combination needed to insulate wire taps and such things. This stretchy stuff is easily torn with the fingers, but makes a watertight and permanent repair in electrical insulation or over electrical repairs where insulation is needed. The plastic tape will withstand high voltages and current even in quite thin layers. Check the tape label to make sure it is not simply a plastic repair tape. These are handy, often attractive and long-lasting, but they are not meant to serve as electrical repair items.

Electrical tape made of vinyl is always stretched as it is applied. It provides a very tight and smooth seal that makes the finished job virtually impervious to water and oil damage as well as electrically safe.

Sponge tapes are available from many companies, including Stanley, Thermwell, Manco, and 3M, and they serve a number of purposes—depending on the type of sponge, the width of the tape, the type of tape backing and the type of adhesive. Most such tapes use a waterproof adhesive with a sponge back varying in width and thickness and sponge material depending on job needs and your desires. Some are used as weatherseals around doors and windows and some are used to keep items from scuffing other surfaces. Others are used as a combination of those two, such as between a camper cab and pickup bed. Other types form seals and buffers at camper cap windows where they look into the pickup window.

Metal tape, with an asphalt adhesive on it, can be used for roof and gutter repairs. Metal tape with other forms of adhesive can be used as underlays on auto body repairs (before the Bondo is placed).

Fig. 8-9. More uses for Flashband. Courtesy of Evode, Inc.

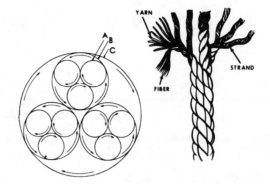

Fig. 8-10. Fiber line make up.

Metal tapes work best when they're not subjected to extremes of heat and cold, as do most all adhesives. Manco makes a monmetallic tape with an asphalt adhesive backing that can even be better than the metallic tapes. The tape is a stretchy and flexible elastomeric butyl rubber, backed with black adhesive.

Velcro is a mechanical fastener in tape form. The material is a hook and loop type of fabric that pulls apart and rejoins readily. From fastening collars on rainwear to working as fastenings on various types of adhesive-backed units for wall uses, to fastening some types of seat covers, to working as tongue closures on hiking boots.

Other types of tape include freezer tapes not affected much by cold down to a few degrees below zero, decorative plastic tapes meant for nothing more than that, clear plastic repair tapes, and pipe-thread sealer tape. This latter style of tape is a Teflon-based material that is wrapped around pipe threads very tightly before you assemble a joint. Leaks are thereby prevented, and the old stand-bys such as pipe dope and white lead are seldom if ever needed.

Rope is a common fastener in any of its forms, from binder twine through heavy Manila hemp and sisal types. In between we have the various plastic types to provide near total water resis-

Fig. 8-11. Three types of fiber line.

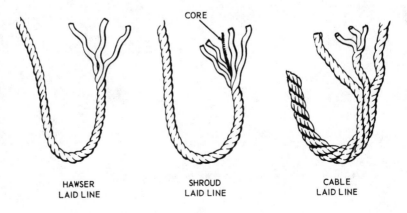

HAWSER LAID LINE

SHROUD LAID LINE

CABLE LAID LINE

MANILA LINE

SOME COMMONLY USED SIZES

	CIRCUMFERENCE	THREAD
	¾ "	6
	1"	9
	1⅛"	12
	1¼"	15
	1½"	21
	1¾"	24
	2"	
	3"	
	4"	
	5"	
	6"	

Fig. 8-12. Common manila rope sizes.

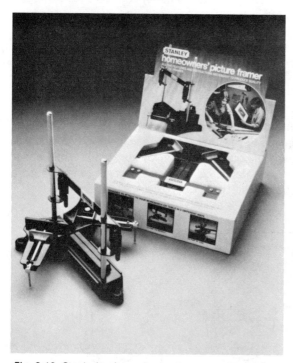

Fig. 8-13. Stanley's picture frame clamps do a number of jobs very well at reasonable cost. Courtesy of The Stanley Works.

Fig. 8-14. This bench hold-down clamp can be invaluable in a well-equipped woodworking shop. Courtesy of Shopsmith, Inc.

Fig. 8-15. Bench end clamps. Courtesy of Shopsmith, Inc.

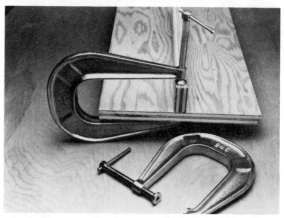

Fig. 8-16. Deep C clamps are exceptionally useful. Courtesy of Shopsmith, Inc.

Fig. 8-17. Corner clamps. Courtesy
of Shopsmith, Inc.

tance. Other types of rope fabrics include polypropylene, nylon,
Dacron, and other types of man-made fibers.

Plaited man-made fibers are almost certain to be lighter and
stronger than the Manila and other natural-fiber ropes. They are
also more solidly resistant to all but sunlight. Constant exposure to
the sun's radiation will destroy polypropylene rope.

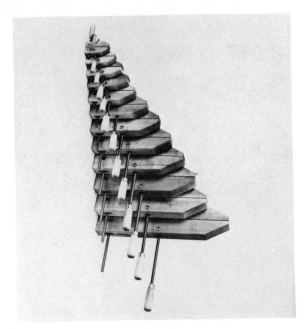

Fig. 8-18. An array of handscrews,
one of the handiest wood clamps
going, with non-marring wood jaws
and adaptability to odd shapes.
Courtesy of Shopsmith, Inc.

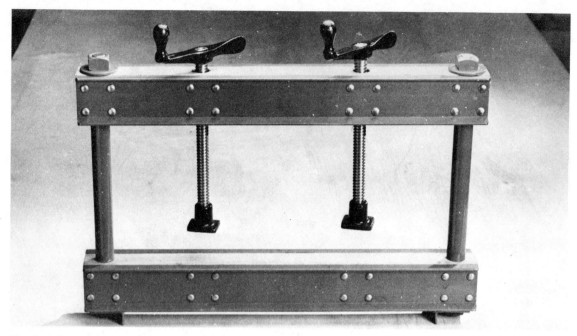

Fig. 8-19. *Press clamp. Courtesy of Shopsmith, Inc.*

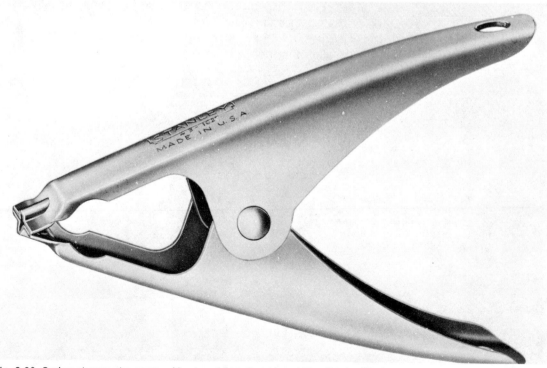

Fig. 8-20. *Spring clamps also come without padding. Courtesy of The Stanley Works.*

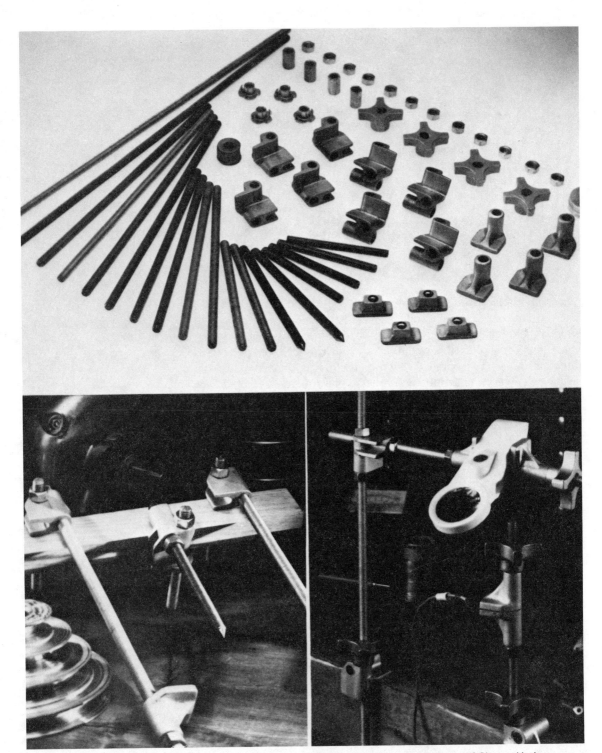

Fig. 8-21. Shopsmith's Maxi-Clamp system offers great versatility at moderate cost. Courtesy of Shopsmith, Inc.

Fig. 8-22. Small, light bar clamps are ideal for gluing up thin sections that might be warped if heavier clamps were used. Courtesy of Shopsmith, Inc.

Plaited—usually eight-strand—rope is considered the modern and best way to make rope. It offers some distinct advantages over braided rope. It's far simpler to splice, and it will rapidly take and hold a temporary splice that would quickly ruin a braided rope. The direction of the yarn twist in plaited rope is of no importance when splicing (it's very important when splicing braided rope). Plaited rope, whether of synthetic or natural fibers, is not as subject to

Fig. 8-23. These oddly shaped clamps do the job of longer clamps by gripping the longer section of the work, saving cost and weight. Courtesy of Shopsmith, Inc.

Fig. 8-24. Bench clamps. Courtesy of Shopsmith, Inc.

Fig. 8-25. The Maxi-Clamp system aids in gluing up irregular shapes, allowing you to design your own clamps as needed. Courtesy of Shopsmith, Inc.

opening and allowing dirt to enter. It tends to "relax" when it is not under stress.

Manila rope is made of select natural fibers. It has been supreme for centuries as the top quality rope for marine and agricultural uses.

Nylon rope was the first challenger to Manila rope. Nylon is

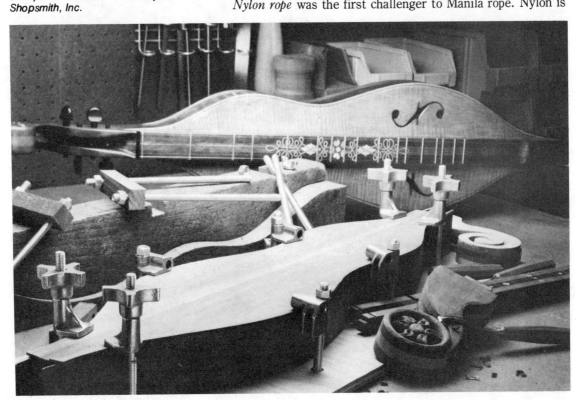

Fig. 8-26. Bar clamps, handscrews and C clamp. Courtesy of Shopsmith, Inc.

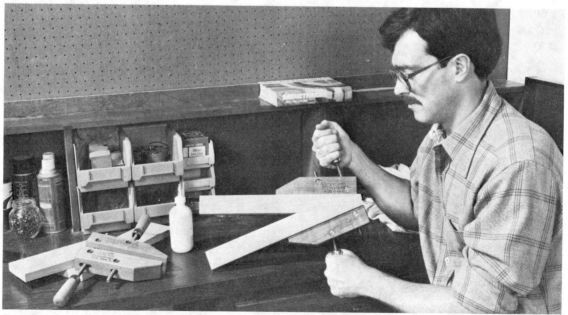

Fig. 8-27. Even hard-to-hold shapes are readily clamped for gluing using handscrews. Courtesy of Shopsmith, Inc.

Fig. 8-28. The array of clamps shown would serve as an excellent midpoint for almost any woodworking ship: web clamp; bench clamps; handscrews; Maxi-Clamp. Courtesy of Shopsmith, Inc.

some two to three times as strong and four to five times as durable as Manila.

Polypropylene rope is very strong, lightweight, and presumably easy to handle. I find it so lightweight and stretchy—and synthetic fibers do not "stretch out" as natural fibers do—it's almost impossible to tie with any success. Splicing is quite easy in braided styles.

Polyethylene rope is even lighter and greasier feeling than polypropylene rope. Again the braided types splice easily.

Polyester (Dacron) rope is my favorite of the synthetics. It's as strong and durable as nylon, it has very low moisture absorption, high resistance to most all forms of wear, and a minimum of stretchiness. It is available in a number of styles. The braided types are easily spliced, as are all braided ropes, while the basic Dacron rope does a better job of maintaining the old granny knot most of us tie than does any other rope other than Manila. Proper knots mean better work from ropes.

Chapter 9

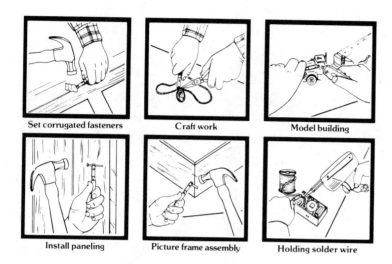

Set corrugated fasteners   Craft work   Model building

Install paneling   Picture frame assembly   Holding solder wire

Miscellaneous Fasteners

There are a wide range of styles and sizes of connectors that form the standard wire nut to types of fasteners crimped right on over insulation and all. See Figs. 9-1 through 9-62. Probably of most importance, besides the wire nuts, are the terminal crimp-on fasteners and the splice fasteners (also in crimp-on styles). Today, solderless terminal lugs see far more use than do the soldered type. They provide decent electrical contact and very high mechanical strength—without presenting you with the common problems of solder joints.

Two of those common problems, cold solder joints—usually from movement of the joint before the hot solder cools—and burned insulation can be of great importance. Terminal lugs today are sold in suitable sizes for just about any terminal you might have to fit. The design of the clamping system to mechanically hold the wire or cable in place depends in large part on the size of the cable and the size of the terminal. Most terminal lugs are crimp-on styles (up to about a size #10). You can fit wire to size #10 without going to special terminal lugs and tools. At this point, you might start to see bolt-down types for the heavy wiring jobs. These offer a bolt through the top of the insert collar that tightens down on the cable—or tightens an internal clamp—to make sure the mechanical strength matches any possible pull that is liable to hit it.

Other varieties of fastener on the terminals are found, with

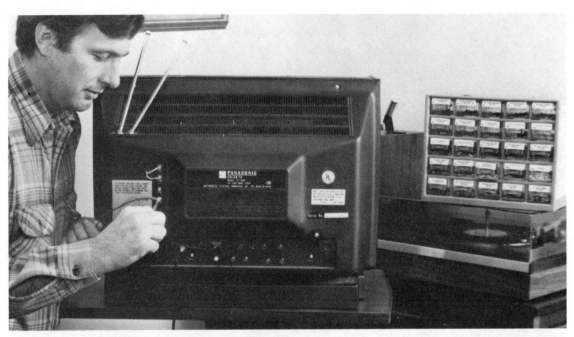

Fig. 9-1. Electrical connectors in use. Courtesy of DRI, Inc.

Fig. 9-2. Electrical connector kit. Courtesy of DRI, Inc.

251

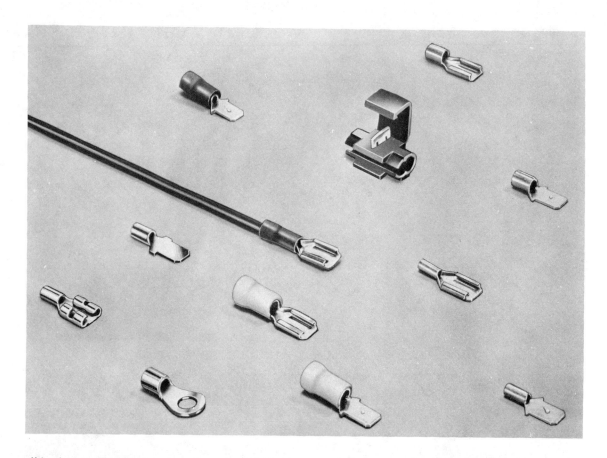

Fig. 9-3. Vaco electrical fasteners.
Courtesy of Vaco.

slide rings over split sleeves and other such devices, and your local electrical supply house should have a selection. For most common uses, the crimp-on styles and the bolt-down styles will be more than sufficient. Terminal lugs are available with uninsulated and insulated shank, up to about a #8 wire size, and sometimes to even larger (smaller numerically) wire and cable sizes.

Crimp-on styles are simply placed in a fairly lightweight pair of special pliers, such as Stanley's Jobmasters, with about ¼ inch of the bared wire inserted into the area to be crimped (the length of wire to be bared will vary with the heaviness of the wire). You then make sure the bared end of the wire is about as far as it will go into the terminal shank, settle everything down in the pliers, and squeeze. The joint is made. It's far more easily than with solder.

Splicing connectors of the crimp-on type are used in the same manner, but they have bared wire entering from both sides so that two bared wires enter the connector—one from each end. Crimping is done in the same manner and the joint is finished (if you've used an insulated connector). If the connector is not insulated, then you must insulate it using at least a single, overlapped, wrap of plastic

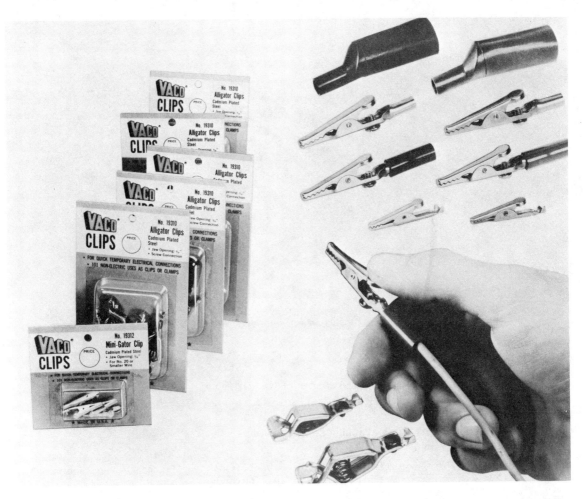

Fig. 9-4. Clip fasteners. Courtesy of
Vaco.

electrician's tape. For wires over #18 in size, I use a double wrap.
For wires over #14 in size, I go to a triple wrap.

Special connectors, such as the three-way Vaconnector, are
used in various applications, including some auto accessories where
you must tap into an already in place wire, on boats and in vans, and
sometimes for trailer wiring hook-ups. Select the spot where you
want the two wires—one continuous and one ending—to meet. Do
not strip the wires. Place the through (uncut) wire in the lower part
of the Vaconnector body and the tapping wire in the upper hole.
Again, do not strip the tapping wire end.

Use a standard pair of pliers to squeeze shut the metal clip top
of the Vaconnector. This drives the clip down through the insulation
on both wires and gives you a good contact. Close the insulating
cover and the job is done.

It sounds far more complex than it is because the actual tap-in
job, once the wires are selected or run, takes about four seconds.

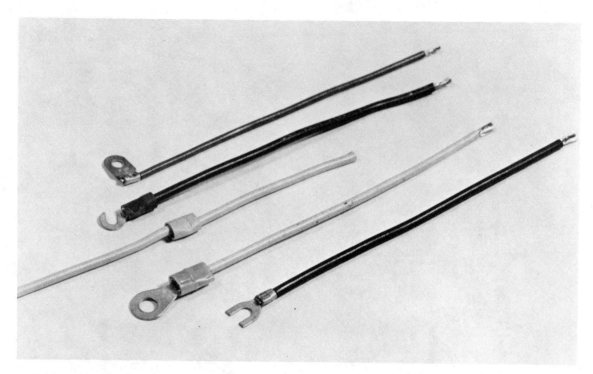

Vaconnectors are available for use with a number of wire sizes.

Other electrical wiring accessories make almost all such jobs a great deal easier these days. Various-length nylon ties with tongues slipping through a head are readily available so that neat and safe wiring placement is far simplified.

Metal fastening hangers and plates of widely varied designs are on the market now. Individually, and in bulk, they're fairly inexpen-

Fig. 9-5. Crimp-on fasteners. Courtesy of Vaco.

Fig. 9-6. Crimp-on terminals. Courtesy of Vaco.

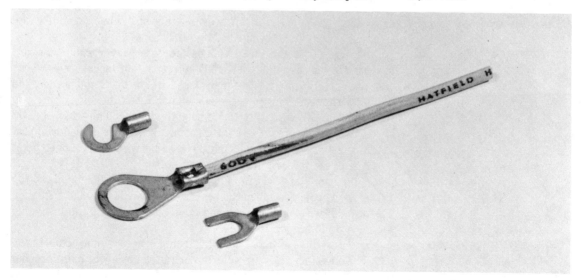

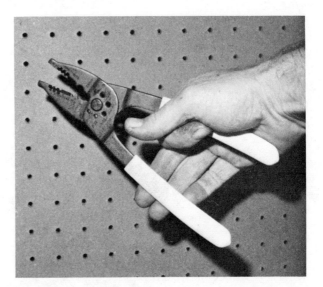

sive. TECO provides specialty nails to make sure the job is done correctly.

Structural wood connectors save a great deal of time and aggravation. They eliminate the need for toenailing at stud bottoms, at joists, and in other places where you can lose time and accuracy when you're not used to the work on a day-in, day-out basis.

Anchor clips are used to tie frame walls into masonry foundations. This eliminates the need for inserting anchor bolts in the foundation as it's built. The clips shape readily and well to most uses, and are made of zinc-coated 16-gauge steel. Upper arms grab the sill plate and the lower members are embedded in the masonry.

Post base clips are another form of wood-to-masonry fastener. They fit over a bolt embedded in a masonry pier or foundation of another type. Use is not limited to houses. These are handy in deck construction where stresses are liable to be above normal. Piers are used under supporting posts, and they can also prove handy in constructing small bridges and many other around-the yard items.

Beam anchors tie a beam into a post and are made, like most of the listed anchors, of 16-gauge steel with zinc or other rust-preventative coatings. Each beam anchor will cover one-half a beam—one side—and post setup. Two are recommended for heavily stressed joints.

Hammer-in bridging of metal is not going to sound like much to someone who has never had to cut and install the bridging for a full floor, but for anyone who has the lack of need for measuring and cutting—along with the lack of need for carting around an apron full of nails during the job—will be more than enough to offset any possible cost increase (which doesn't exist once you check labor needs for the two methods).

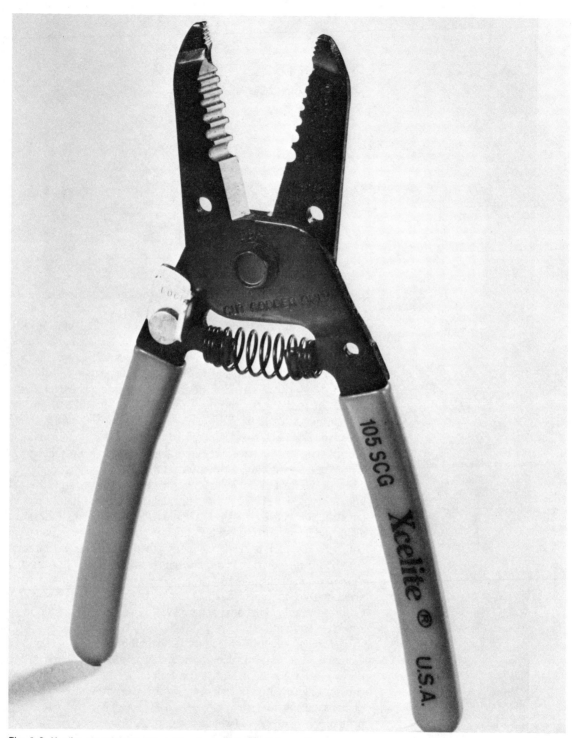

Fig. 9-8. Xcelite electrician's pliers. Courtesy of Xcelite/Cooper Group.

256

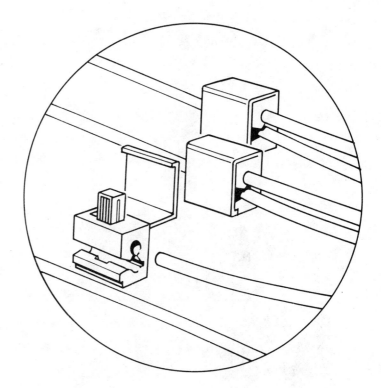

Fig. 9-9. Three-way Vaconnector.
Courtesy of Vaco.

Sizes for 16-inch on-center and 24-inch on-center installations are available. One end is straight—but pointed or sawtoothed—and the other end is bent. To install the bridging, simply nail the straight end into one joist and then push the bent end up until it's ready to be hammered into the opposite joist. Hammer it in and you're done. Minor variations in size make no difference at all.

If you are working with normal wood bridging materials for cut-in bridging, you must first get your measurements for *each* joist-to-joist difference. Then cut 45-degree angles at the appropriate distance on the bridging material (usually 1 × 4). This requires a second measurement to make sure the angle doesn't throw you off. The bridging pieces must be longer than the distance between joists, of course, because they're cut in at an angle. Two bridging pieces per spot being bridged are used, crossing each other (this is also done with metal bridging), and 2-, 8-, or 10-penny nails must be installed at each end of each piece. If nothing else, you've got eight nail-driving operations replaced by only four with the metal strips.

Framing clips or anchors are odd-looking little things of 18-gauge metal, usually zinc coated, with about a dozen nail holes. The slotted portions can be left straight or bent to suit to right or left or

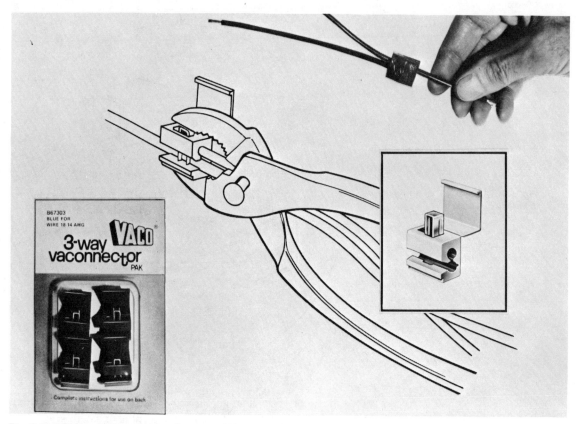

Fig. 9-10. Three-way Vaconnector. Courtesy of Vaco.

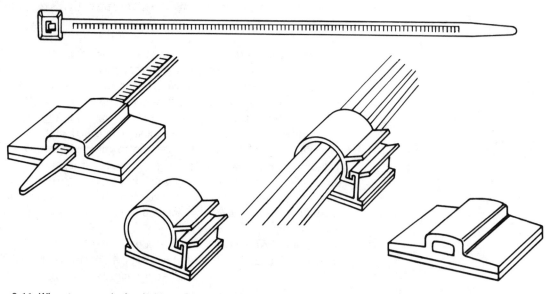

Fig. 9-11. Wire straps and other holders. Courtesy of Vaco.

ABLE ANCHOR sill plate anchor

WALL BRACING

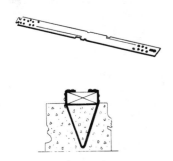

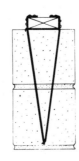

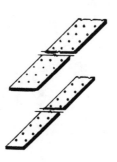

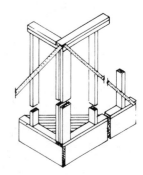

Fig. 9-12. Sill plate anchor. Courtesy of TECO Products & Testing Corp.

at angles other than 90 degrees. Most commonly, these are used to eliminate toenailing of studs, but they can also be used in other ways that depend on your needs at the time.

Twist or *storm anchors* are used to eliminate possible windlift problems with roof decking. They anchor the roof rafters firmly to the top plates of the stud walls. Made of 18-gauge, zinc-coated steel, these anchors take far less time than you would spend using conventional methods of anchoring and they are stronger.

Joist hangers are a joy to use and they are a time-saver in any form of frame construction where joists must be hung (whether it's a house you're putting up or a deck). Nailing and positioning tabs on these 16-gauge steel hangers make lining joists up quite simple. No movement is allowed when nailing is done (toenailing often drives a joist off its mark). There's also no need for a ledger board. The resulting notch must then be cut in the bottom of each joist to fit over the board so that a great deal of installation time and aggravation is saved.

POST CAPS

POST ANCHOR (base)

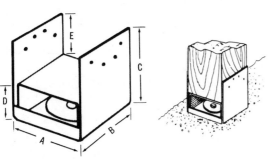

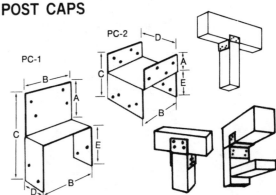

Fig. 9-13. Post anchor. Courtesy of TECO Products & Testing Corp.

Fig. 9-14. Post caps. Courtesy of TECO Products & Testing Corp.

FAS-LOK metal bridging

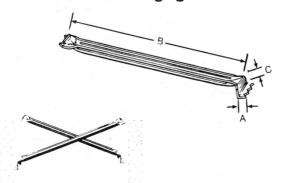

Fig. 9-15. Metal bridging. Courtesy of TECO Products & Testing Corp.

TECO-U-GRIP joist and beam hangers

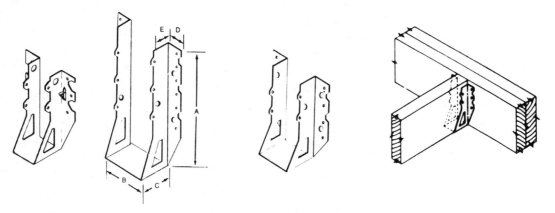

Fig. 9-16. Joist and beam hangers. Courtesy of TECO Products & Testing Corp.

H-CLIP plywood supports   SHIMS

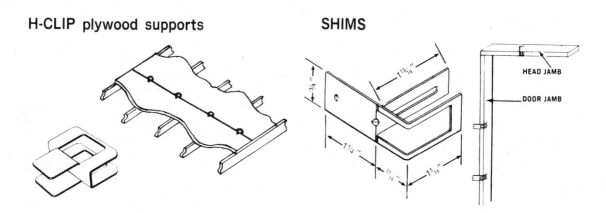

Fig. 9-17. H-clip plywood supports. Courtesy of TECO Products & Testing Corp.

260

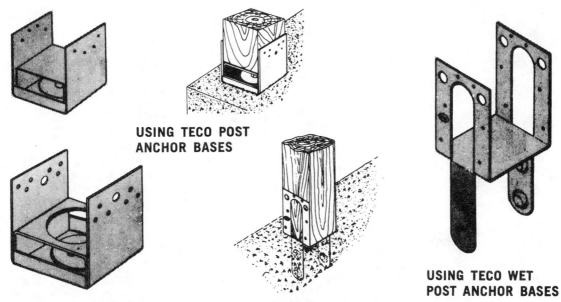

**USING TECO POST
ANCHOR BASES**

**USING TECO WET
POST ANCHOR BASES**

Fig. 9-18. Anchor bases. Courtesy of TECO Products & Testing Corp.

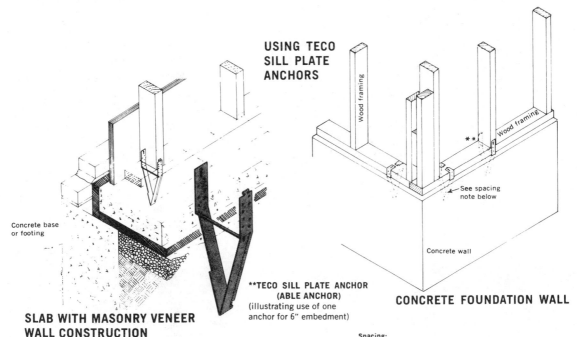

**USING TECO
SILL PLATE
ANCHORS**

Wood framing

Wood framing

**See spacing
note below

Concrete base
or footing

Concrete wall

**TECO SILL PLATE ANCHOR
(ABLE ANCHOR)**
(illustrating use of one
anchor for 6″ embedment)

**SLAB WITH MASONRY VENEER
WALL CONSTRUCTION**

CONCRETE FOUNDATION WALL

Note:
Proper installation in all cases assumes use of 8d common
nails, with at least 4 nail holes filled on each side. In case of
plate, with end of anchor bent over onto top of plate, at least
2 nails are driven into each side of plate and at least 2 nails
into each bent down end.

Spacing:
Spacing of anchors is the same as, and need not exceed, that
for conventional anchor bolts. Maximum spacing shall be 8′
o.c. with not less than two anchors in each sill piece. End
anchors shall not be more than 12″ from the end of the piece.
Where earthquake design is required, maximum spacing shall
be 6′ o.c.

Fig. 9-19. Sill plate anchors. Courtesy of TECO Products & Testing Corp.

Several methods of framing: FLOOR LEVEL

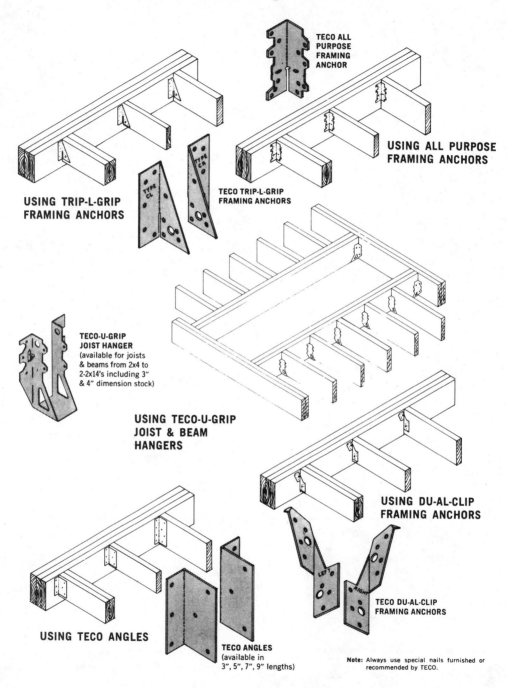

TECO ALL PURPOSE FRAMING ANCHOR

USING ALL PURPOSE FRAMING ANCHORS

USING TRIP-L-GRIP FRAMING ANCHORS

TYPE CL

TYPE CR

TECO TRIP-L-GRIP FRAMING ANCHORS

TECO-U-GRIP JOIST HANGER (available for joists & beams from 2x4 to 2-2x14's including 3" & 4" dimension stock)

USING TECO-U-GRIP JOIST & BEAM HANGERS

USING DU-AL-CLIP FRAMING ANCHORS

LEFT

RIGHT

TECO DU-AL-CLIP FRAMING ANCHORS

USING TECO ANGLES

TECO ANGLES (available in 3", 5", 7", 9" lengths)

Note: Always use special nails furnished or recommended by TECO.

Fig. 9-20. Floor framing anchors. Courtesy of TECO Products & Testing Corp.

Several methods of framing: WALL CONNECTIONS

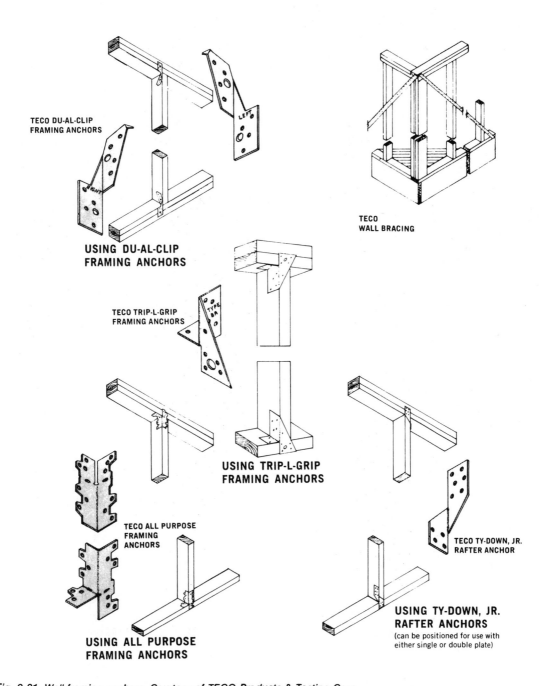

TECO DU-AL-CLIP
FRAMING ANCHORS

USING DU-AL-CLIP
FRAMING ANCHORS

TECO
WALL BRACING

TECO TRIP-L-GRIP
FRAMING ANCHORS

USING TRIP-L-GRIP
FRAMING ANCHORS

TECO ALL PURPOSE
FRAMING
ANCHORS

TECO TY-DOWN, JR.
RAFTER ANCHOR

USING ALL PURPOSE
FRAMING ANCHORS

USING TY-DOWN, JR.
RAFTER ANCHORS
(can be positioned for use with
either single or double plate)

Fig. 9-21. Wall framing anchors. Courtesy of TECO Products & Testing Corp.

Several methods of framing: ROOF ANCHORAGE

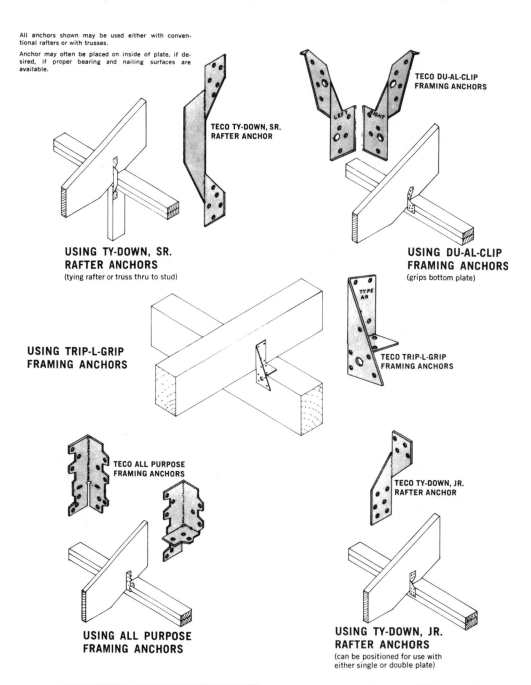

All anchors shown may be used either with conventional rafters or with trusses.

Anchor may often be placed on inside of plate, if desired, if proper bearing and nailing surfaces are available.

TECO TY-DOWN, SR. RAFTER ANCHOR

TECO DU-AL-CLIP FRAMING ANCHORS

LEFT   RIGHT

**USING TY-DOWN, SR.
RAFTER ANCHORS**
(tying rafter or truss thru to stud)

**USING DU-AL-CLIP
FRAMING ANCHORS**
(grips bottom plate)

**USING TRIP-L-GRIP
FRAMING ANCHORS**

TYPE AR

TECO TRIP-L-GRIP FRAMING ANCHORS

TECO ALL PURPOSE FRAMING ANCHORS

TECO TY-DOWN, JR. RAFTER ANCHOR

**USING ALL PURPOSE
FRAMING ANCHORS**

**USING TY-DOWN, JR.
RAFTER ANCHORS**
(can be positioned for use with
either single or double plate)

Fig. 9-22. Roof anchors. Courtesy of TECO Products & Testing Corp.

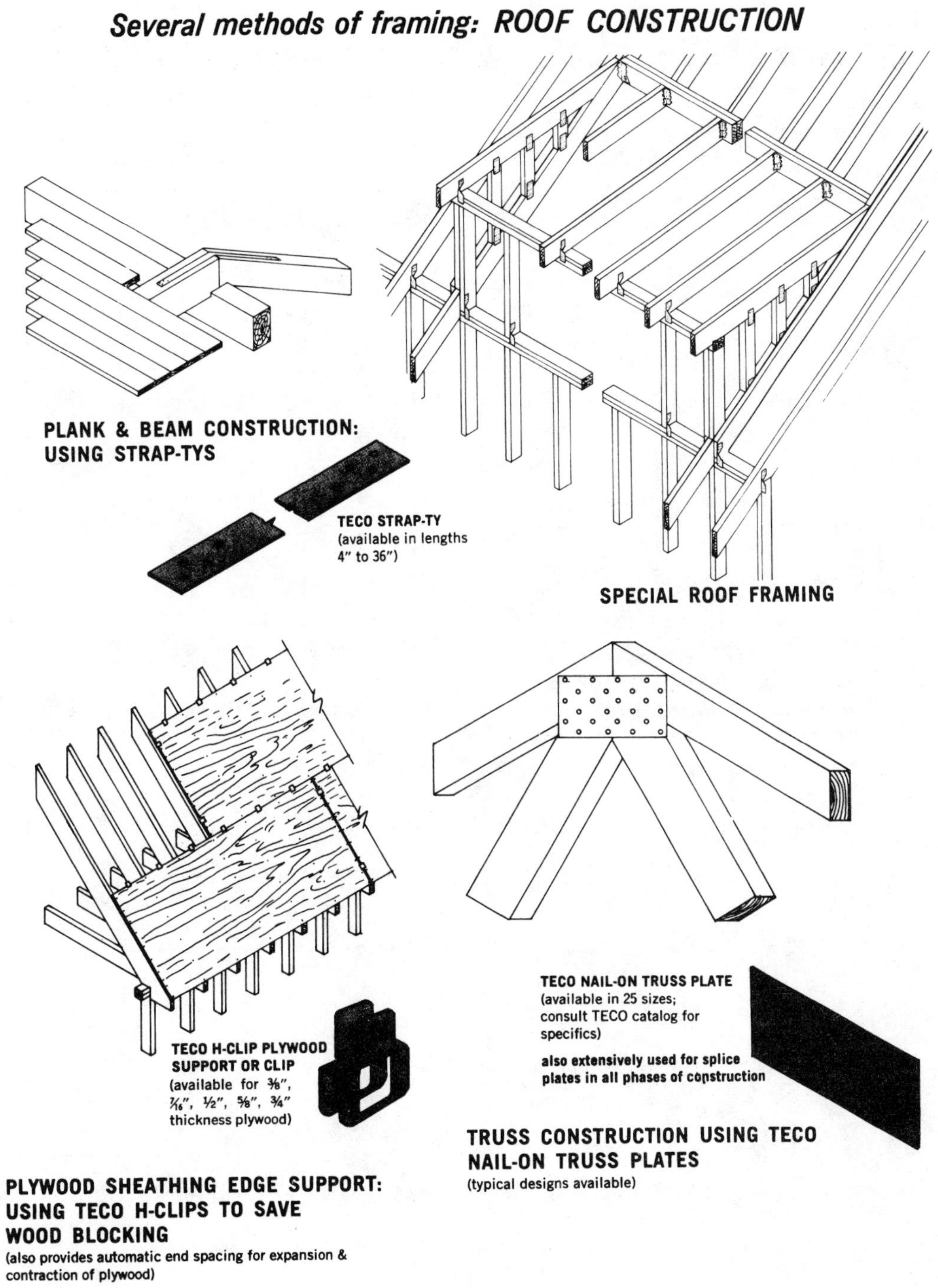

Several methods of framing: ROOF CONSTRUCTION

PLANK & BEAM CONSTRUCTION: USING STRAP-TYS

TECO STRAP-TY
(available in lengths
4" to 36")

SPECIAL ROOF FRAMING

TECO H-CLIP PLYWOOD SUPPORT OR CLIP
(available for ⅜",
⁷⁄₁₆", ½", ⅝", ¾"
thickness plywood)

PLYWOOD SHEATHING EDGE SUPPORT: USING TECO H-CLIPS TO SAVE WOOD BLOCKING
(also provides automatic end spacing for expansion &
contraction of plywood)

TECO NAIL-ON TRUSS PLATE
(available in 25 sizes;
consult TECO catalog for
specifics)

also extensively used for splice
plates in all phases of construction

TRUSS CONSTRUCTION USING TECO NAIL-ON TRUSS PLATES
(typical designs available)

Fig. 9-23. Roof framing aids. Courtesy of TECO Products & Testing Corp.

Several methods of framing: MISCELLANEOUS APPLICATIONS

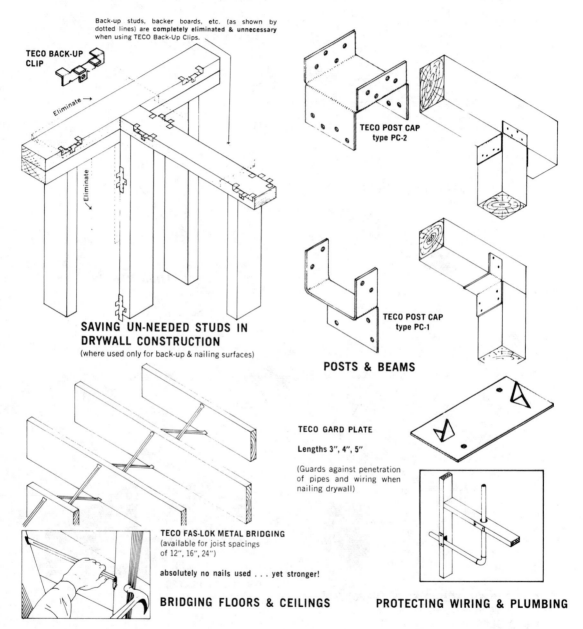

Back-up studs, backer boards, etc. (as shown by dotted lines) are **completely eliminated & unnecessary** when using TECO Back-Up Clips.

TECO BACK-UP CLIP

Eliminate →

← Eliminate

SAVING UN-NEEDED STUDS IN DRYWALL CONSTRUCTION
(where used only for back-up & nailing surfaces)

TECO POST CAP type PC-2

TECO POST CAP type PC-1

POSTS & BEAMS

TECO GARD PLATE

Lengths 3″, 4″, 5″

(Guards against penetration of pipes and wiring when nailing drywall)

TECO FAS-LOK METAL BRIDGING
(available for joist spacings of 12″, 16″, 24″)

absolutely no nails used . . . yet stronger!

BRIDGING FLOORS & CEILINGS

PROTECTING WIRING & PLUMBING

Fig. 9-24. Miscellaneous framing aids. Courtesy of TECO Products & Testing Corp.

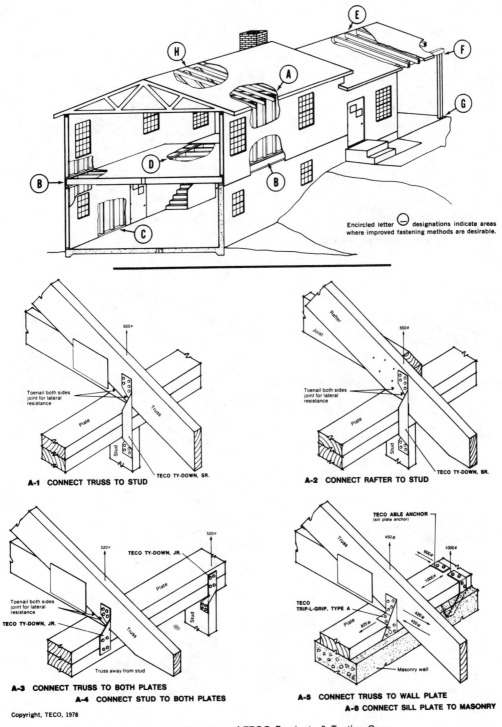

Fig. 9-25. Storm resistant construction aids. Courtesy of TECO Products & Testing Corp.

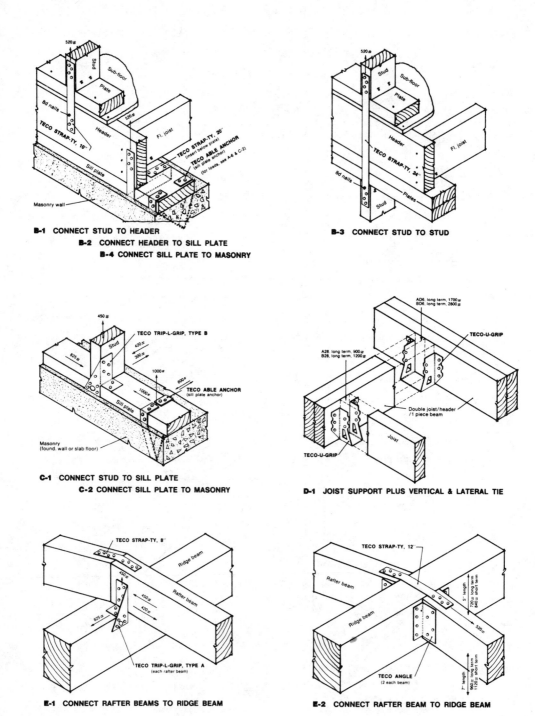

B-1 CONNECT STUD TO HEADER
B-2 CONNECT HEADER TO SILL PLATE
B-4 CONNECT SILL PLATE TO MASONRY

B-3 CONNECT STUD TO STUD

C-1 CONNECT STUD TO SILL PLATE
C-2 CONNECT SILL PLATE TO MASONRY

D-1 JOIST SUPPORT PLUS VERTICAL & LATERAL TIE

E-1 CONNECT RAFTER BEAMS TO RIDGE BEAM

E-2 CONNECT RAFTER BEAM TO RIDGE BEAM

Fig. 9-26. Storm resistant construction aids. Courtesy of TECO Products & Testing Corp.

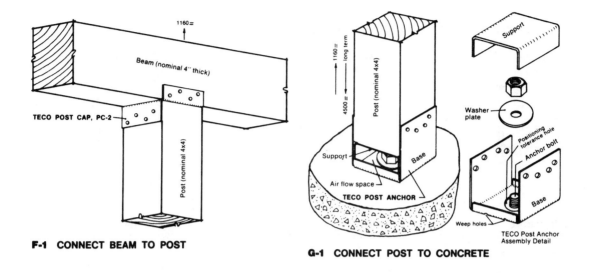

F-1 CONNECT BEAM TO POST

G-1 CONNECT POST TO CONCRETE

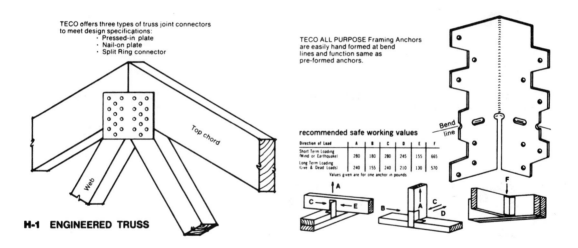

H-1 ENGINEERED TRUSS

TECO offers three types of truss joint connectors to meet design specifications:
· Pressed-in plate
· Nail-on plate
· Split Ring connector

TECO ALL PURPOSE Framing Anchors are easily hand formed at bend lines and function same as pre-formed anchors.

recommended safe working values

Direction of Load	A	B	C	D	E	F
Short Term Loading (Wind or Earthquake)	280	180	280	245	155	665
Long Term Loading (Live & Dead Loads)	240	155	240	210	130	570

Values given are for one anchor in pounds

Fig. 9-27. Storm-resistant construction aids. Courtesy of TECO Products & Testing Corp.

The joist hangers are sized for different uses from as small as 2 × 4 on up to doubled 2 × 6 and larger joists. The joist hanger is simply nailed in place at the mark. The joist is cut to length, set in place, and then nailed to the hanger.

You can then run one toenail down from the joist into the beam, but all this really does in most cases is interfere with floor nailing. Special nails are supplied with the better joist hangers (such as those made by TECO).

Panel clips are useful when you lay plywood roof decking, subflooring, and so on. The clips save you the trouble of aligning panel edges with joists, rafters, etc., and are generally acceptable

APPLICATION DETERMINES KIND OF HINGE

Full mortise
Wood Doors, Wood Jamb

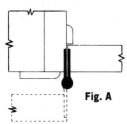

Fig. A

Full mortise
Wood Doors, Hollow Metal Jamb

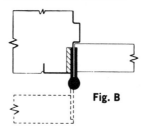

Fig. B

Full mortise
Hollow Metal Doors, Hollow Metal Jamb

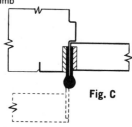

Fig. C

Half mortise
Hollow Metal Doors, Channel Iron Jamb

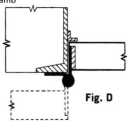

Fig. D

Half surface
Wood Doors, Wood Jamb

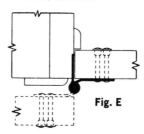

Fig. E

Half surface
Mineral Core Doors, Hollow Metal Jamb

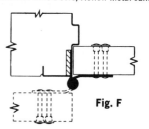

Fig. F

Full surface
Mineral Core Doors, Channel Iron Jamb

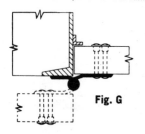

Fig. G

Full surface
Hollow Metal Doors, Channel Iron Jamb

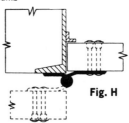

Fig. H

"Swing Clear" full mortise
Wood Doors, Hollow Metal Jamb

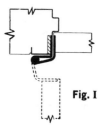

Fig. I

"Swing Clear" half mortise
Wood Doors, Channel Iron Jamb

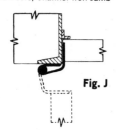

Fig. J

"Swing Clear" half surface
Mineral Core Doors, Hollow Metal Jamb

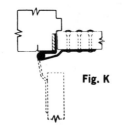

Fig. K

"Swing Clear" full surface
Mineral Core Doors, Channel Iron Jamb

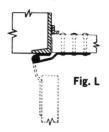

Fig. L

Fig. 9-28. Hinge applications. Courtesy of The Stanley Works.

Doors opening in

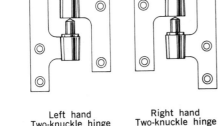

Left hand door
takes
left hand hinges

Right hand door
takes
right hand hinges

Doors opening out
Including closet, cupboard
and bookcase doors

Right hand reverse door
takes
left hand hinges

Left hand reverse door
takes
right hand hinges

Left hand
Two-knuckle hinge

Right hand
Two-knuckle hinge

- The hand of a butt hinge is determined from the outside of the door to which it is applied

- The outside of a cupboard, bookcase or closet door is the room side. For other doors, the outside is usually the "push" or jamb side

- If standing outside of a door which opens **from** you, to the right, it takes right hand butt hinges. If to the left, it takes left hand butt hinges

- If standing outside of a door which opens **toward** you, to the right, it takes left hand butt hinges. If to the left, it takes right hand butt hinges

- Reversed bevel doors are doors opening out

- It will be observed that this method of determining the hand of loose joint cabinet hinges is exactly the opposite from the furniture manufacturers' standard rule

- To determine the hand of a loose joint hinge: open the hinge with its face toward you. If the knuckle of the right leaf is at the bottom, it is a right hand hinge. If the knuckle of the left leaf is at the bottom, it is a left hand hinge

In ordering butt hinges that are not reversible, the hand must always be specified

Fig. 9-29. Determining hinge hand. Courtesy of The Stanley Works.

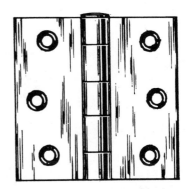

Fig. 9-30. Butt hinge. Courtesy of The Stanley Works.

Fig. 9-31. Butt hinge. Courtesy of The Stanley Works.

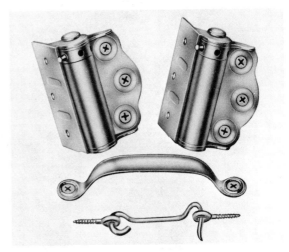

Fig. 9-32. Butt hinges for screen door use. Courtesy of The Stanley Works.

271

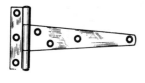 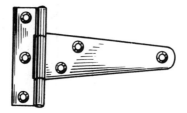 

as structurally as sound. I don't like them. I'd much rather have my panels meet—even if extra cutting is required—on the framing members.

There is a wide variety of other anchor styles such as drywall anchors, jamb anchors, truss clips, top plate ties, and so on. These are essentially similar in use, if not looks, to standard framing anchors. They are far heavier for use in much heavier construction than most of us are used to dealing with.

Fence brackets are something I like a great deal. Board fences in any design are a lot of work to construct. The more complex designs can drive you wild with on-site cutting and nailing. Fence

Fig. 9-33. Strap hinges. Courtesy of The Stanley Works.

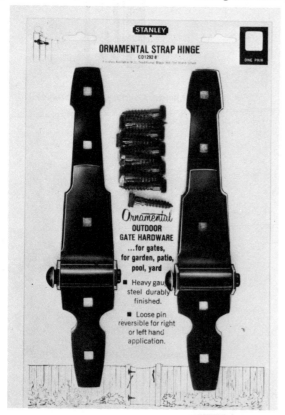

Fig. 9-34. Decorative strap hinge. Courtesy of The Stanley Works.

Fig. 9-35. Mortar mount strap hinge. The flat bar is made to mount in the mortar joint. Courtesy of The Stanley Works.

brackets are nothing more than open-top and closed-bottom clips that nail to a fence post (or bottom or top rail if you're building a louvered fence) to allow you to slip the boards in easily.

The slip-in feature offers some gratifying advantages over nailing boards in place. If a board is broken, it can be easily replaced. The replacement, after cutting the board to length, takes about 10 seconds. Pulling and replacing a nailed board can be a chore that causes the collapse of entire sections of fence.

An additional benefit of slip-out construction is the ease with which large fence sections can be opened to allow passage of larger than normal machinery. Most farm fencing offers gates at least 8 feet wide on up to a dozen feet on single-side gates. Most residential gates are under 4-feet wide. Being able to get, for example, a

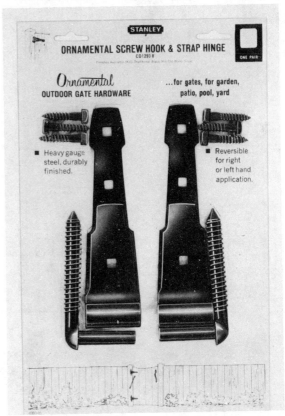

Fig. 9-36. Screw hook strap hinge. Courtesy of The Stanley Works.

backhoe in place might prove a boon at home at times. In addition the fence brackets allow you to easily adapt to any fence design, either vertical or horizontal, and save a great deal of cutting and general work. When I use these on vertical board fences, I like to partially drive a 6-penny nail at the open side of each bracket. If that's done, children, pets, and other low-flying objects cannot jar loose the boards.

Fig. 9-37. Hinge and pull variety. Courtesy of The Stanley Works.

Fig. 9-38. Thumb catch latch. Courtesy of The Stanley Works.

Hinges are used in closed box construction for furniture as well as on doors to homes and in other areas of the home.

Hinges come in two styles. Most lightweight hinges for cabinetry and such work are fixed pin styles. That means you've essentially got a one piece hinge. This is no problem if a door is unlikely to have to be removed, but in residences and other areas where furniture and appliances must be moved with some frequency, fixed-pin hinges are not a good idea. You'll find it's far simpler if you have a loose pin hinge where you don't have to pull and then reinstall from four to eight hinge screws.

Butt hinges fit between the door edge and jamb so that only the hinge pin is exposed on the inside of the door. A *full-mortised butt hinge* is mortised both into the door and into the jamb, and is about

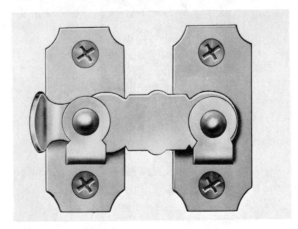

Fig. 9-39. Drop catch. Courtesy of The Stanley Works.

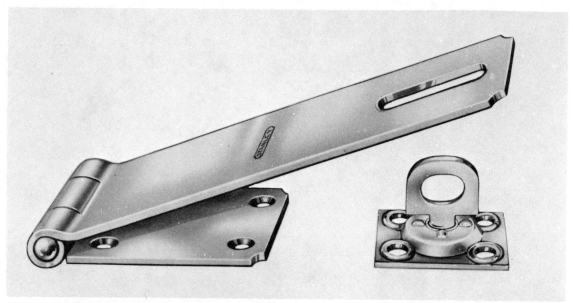

Fig. 9-40. Hasp and staple. Courtesy of The Stanley Works.

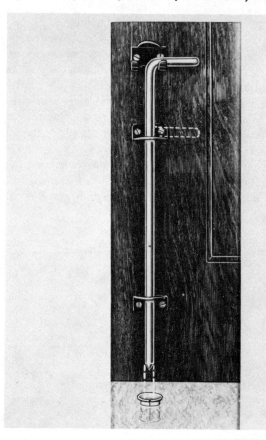

Fig. 9-41. Cane bolt. Courtesy of The Stanley Works.

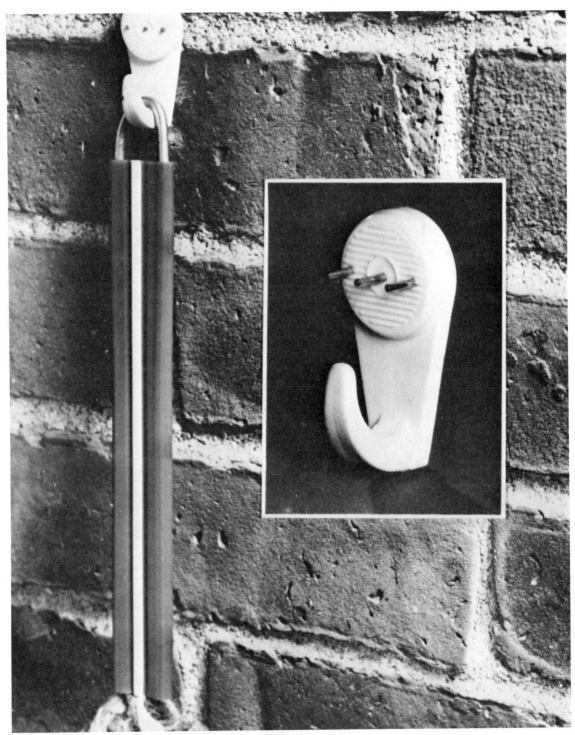

Fig. 9-42. Masonry hanger. Courtesy of The Stanley Works.

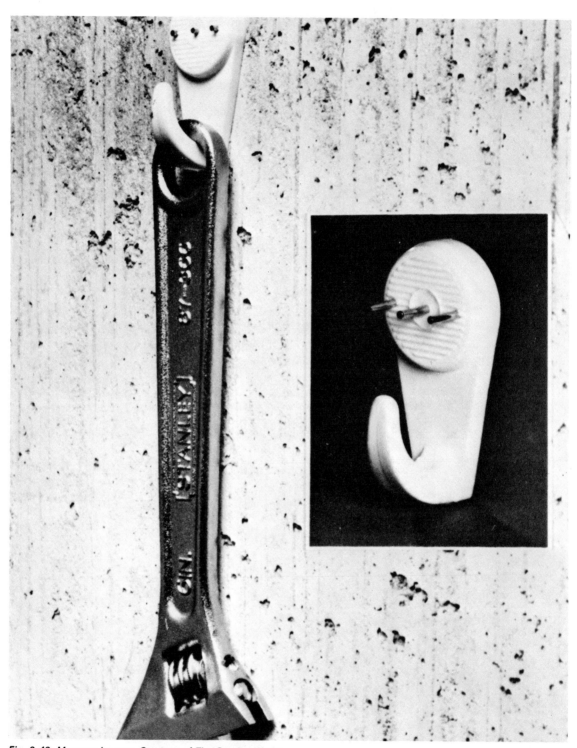

Fig. 9-43. Masonry hanger. Courtesy of The Stanley Works.

Fig. 9-44. Self-stick hanger. Courtesy of *The Stanley Works*.

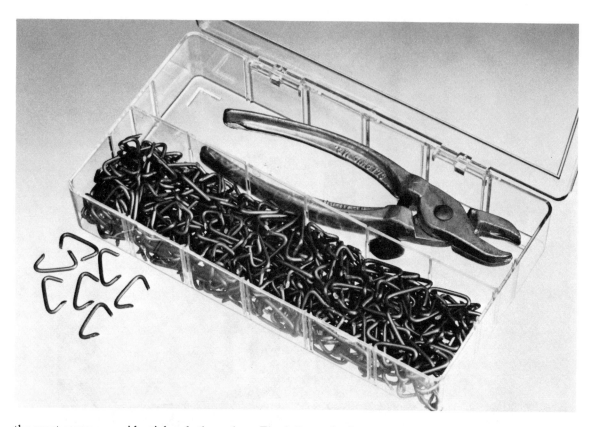

Fig. 9-45. Hog rings Courtesy of DRI, Inc.

the most common residential style these days. The full-mortise butt hinge is made so that both hinge surfaces meet flat when the door is closed. Some of the better models have ball or other kinds of bearings around the pin to add to hinge life.

Full-surface hinges mount with both hinge leaves on the surface of the door and jamb, with the entire hinge exposed when the door is shut. Obviously, these are not worth beans if mounted on the exterior of an exterior door, at least from any security standpoint. They can be modified so that security is improved by using non-removable screws or by running headless bolts all the way through the door.

Half-mortise hinges mount with one hinge leaf mortised and the other applied to the surface. In most cases, the mortised leaf will be on the door jamb and not the door itself.

Pivot hinges mount at door tops and bottoms instead of along the door butt or edge. Only a small bit of metal or other hinge material is in view, and the use is common where doors are not meant to be noticed much or where doors must swing both ways.

Strap hinges are for surface application and offer a rather strange choice. They are either totally utilitarian in appearance (usually zinc plated) and fairly ugly or they are done in a decorative

Fig. 9-46. Right: Snaps. Courtesy of DRI, Inc.

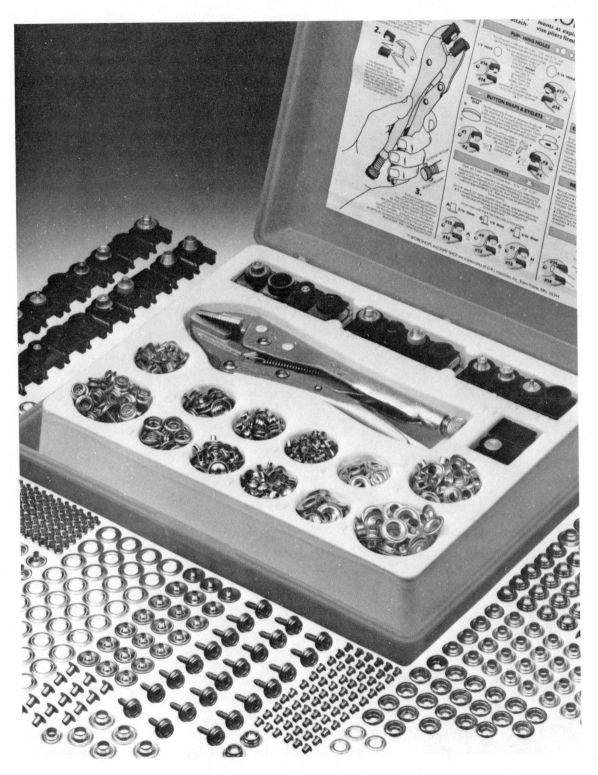

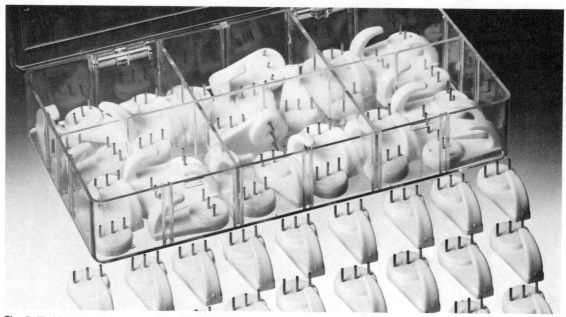

Fig. 9-47. Masonry hangers. Courtesy of DRI, Inc.

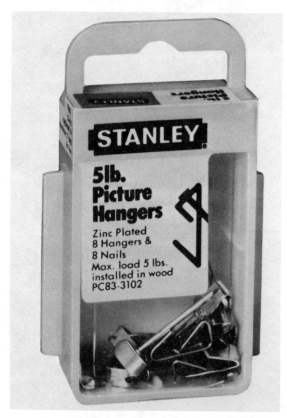

Fig. 9-48. Picture hanger. Courtesy
of The Stanley Works.

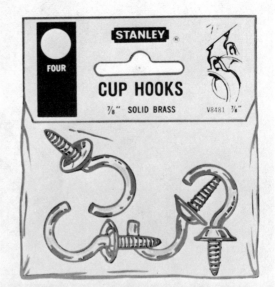

Fig. 9-49. Cup hooks. Courtesy of The Stanley Works.

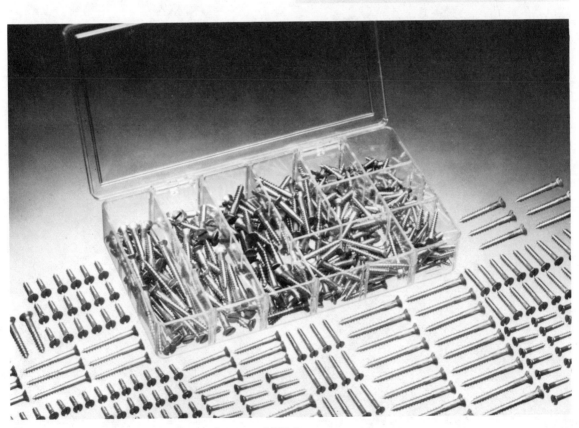

Fig. 9-50. An assortment of wood screws. Courtesy of DRI, Inc.

Fig. 9-51. Masonry nails are wedged shaped. Courtesy of The Vaughan & Bushnell.

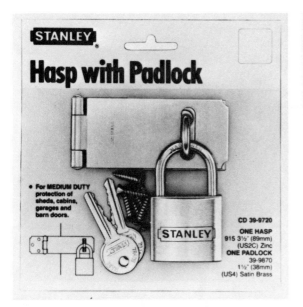

Fig. 9-52. A hasp with padlock. Courtesy of The Stanley Works.

Fig. 9-53. Transparent Weatherproof tape. Courtesy of The Stanley Works.

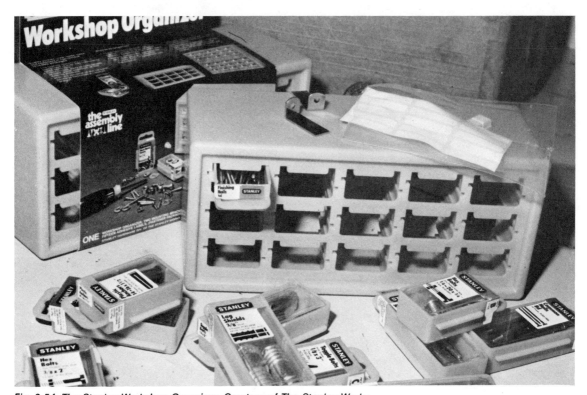

Fig. 9-54. The Stanley Workshop Organizer. Courtesy of The Stanley Works.

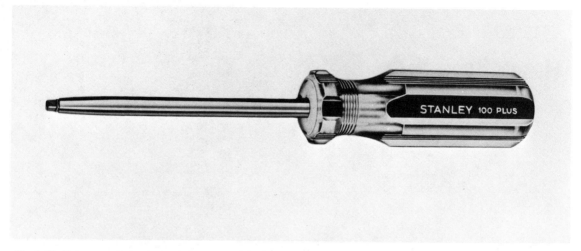

Fig. 9-55. Robertson tip screwdriver. Courtesy of The Stanley Works.

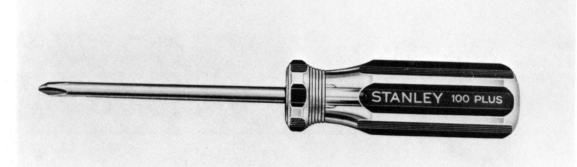

Fig. 9-56. Phillips screwdriver. Courtesy of The Stanley Works.

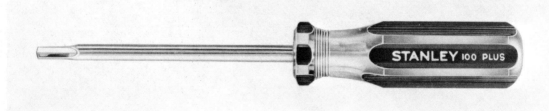

Fig. 9-57. Clutch-tip screwdriver. Courtesy of The Stanley Works.

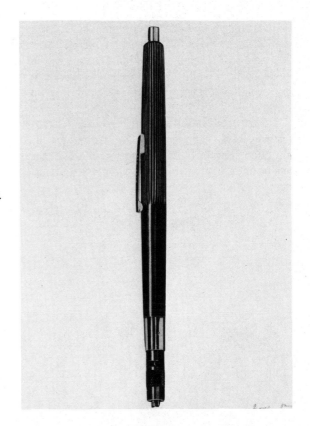

Fig. 9-58. A starter screwdriver.
Courtesy of The Stanley Works.

form (usually what's called Colonial style, in black finish) and just about as heavy duty as the utilitarian models.

Strap hinges install easily and are very popular. Variations on the theme include mortar mounting, screw-in, and other jamb sections for so-called gate hinges.

Piano hinges can run up to and past 6 feet in length and are generally for lightweight to medium-weight uses along toolbox and lifting countertop edges as well as on other forms of box storage uses. They're frequently used in cabinetry where a problem with warping might occur if lighter, shorter hinges are used.

Catches are made in friction, roller spring, magnetic, elbow, bullet, and touch models. For most uses, the friction and magnet styles are probably the most popular, but take a careful look at available models of other designs.

Friction catches hold by pressure of the catch portion in the striker. Magnetic catches usually have a magnet on the door and a metal plate on the strike. Roller spring catches are really a form of friction catch with a roller used instead of a spring style catch. Bullet catches have a spring-loaded bullet that pops into a small depression. Elbow catches have an elbow over which a hook fits. Touch catches operate by touch; no other pressure is required to release the catch and allow the door to open.

Hasps and staples are a form of catch combined with a hinge. These are secured with a padlock, in most instances, but can just have a peg dropped through the staple. There are some versions that have a turning staple that serves as a secure catch, without locking the hasp.

Gate latches are usually versions of other forms of latches simply increased in size and weight. Cane bolts look like walking canes. They slide down or up to hold doors open, gates open, or to fit into slots or other devices to make two-piece units operate as a single piece.

Fig. 9-59. Torx screwdrivers. Courtesy of Vaco.

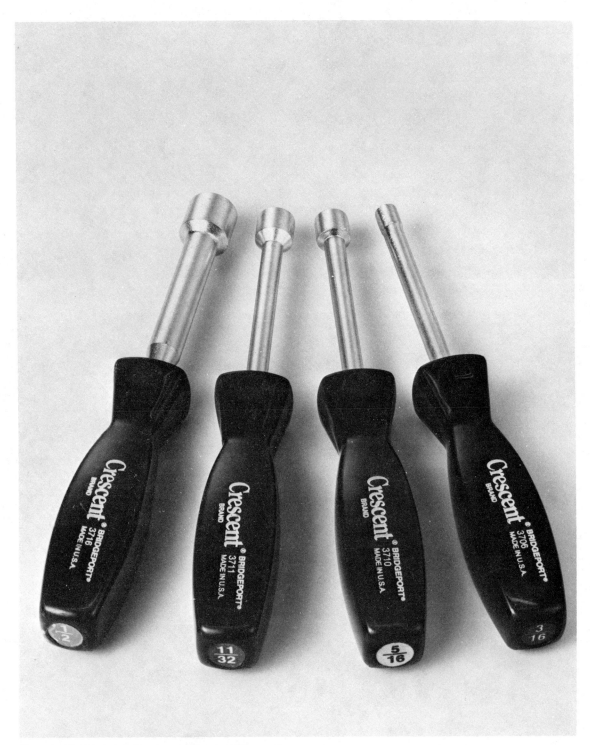

Fig. 9-60. Nutdriver set. Courtesy of Crescent/Cooper Group.

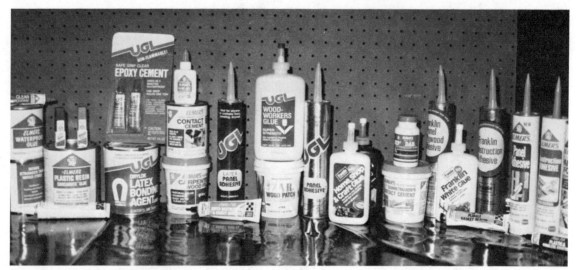

Fig. 9-61. Do-it-yourself gluing products.

Fig. 9-62. UGL's woodworker's glue.

Index

OTHER POPULAR TAB BOOKS OF INTEREST